Steady State Convection

Joseph M. Kats

IrkYork publisher

2019

Contents

NOTATIONS:

$T, T_c, T_h, T_e, T_{cnd}, \Theta$ - temperatures,

P, P_e, p - pressures,

ρ, ρ_0 - mass densities,

e, s, ρ_0 - energy, entropy, density; all are per unit mass,

κ - thermal conductivity,

c - specific heat capacity,

$\chi = \kappa/(c\rho_0)$ - thermal diffusivity

η - dynamic viscosity,

$v = \eta/\rho_0$ - kinematic viscosity,

α - thermal expansion coefficient,

g - gravitational acceleration,

d - depth of liquid layer,

t - time,

$Ra = g\alpha\Theta d^3/(v\chi)$ - Rayleigh number,

$Pr = v/\chi$ - Prandtl number,

Nu – Nusselt number,

L - the longest side of the container,

$\Gamma = L/d$ - the aspect ratio,

$\lambda = 2\pi/k$ – diameter of the convection cell,

k - wave number,

N - number of the vertical nodes,

M - number of the horizontal nodes,

h – the distance between two adjacent nodes.

PREFACE

Convection is the motion of a liquid or a gas due to heating from below. It is the part of hydrodynamics which studies the general motion of fluids. Convection plays a significant role in the Earth's mantle, moving tectonic plates and creating earthquakes. Convection is significant in ocean currents as well as in global air circulation and in meteorology. Convection influences processes in the Sun and stars.

As any physical discipline, convection is researched experimentally and theoretically. This book is about theoretical convection, where some important questions have remained unanswered for a long time. In this book the answers will be given to two questions.

The first question is whether or not the steady state regime of convection depends on initial conditions at which convection started.

The second question is whether the size of convection cells increases or decreases when the temperature of the heating is rising.

Along with vector notations I use tensor representations $\partial_i = \partial x_i$, $\partial_i V_k = \partial V_k/\partial x_i$ and summation is assumed over repeating indices. All definite integrals over volume Ω will be shown as $\int ... d\Omega$.

This book consists of chapters that consist of sections.
The numbering of equations is its own in each section including section number. (3.2) means equation 2 in section 3 in the given chapter. (II.3.2) means equation 2 in section 3 in Chapter II. The same system is used for numbering figures. The only exceptions are in the Introduction where equations are numbered as (*), (**) and figure is numbered as Fig.1.

ACKNOWLEDGEMENTS

I am grateful to Professor E.L. Koschmieder for the fruitful discussions and for his inspiring book.

Many thanks to Mr. J.S. Krasikov for his help in programming, for his support and interest in my work.

I am grateful to my daughter Irina Kats for her help in the editing of this book.

I am grateful to my wife Aria Kats for her continued support and advice.

I am thankful to the colleagues of the Solar Department of the Institute of Solar-Terrestrial Physics in Irkutsk, Russia, who made many useful remarks regarding my talk at their seminar and especially to Dr. V.I. Skomorovsky for his help in researching the literature sources on convection.

INTRODUCTION

Convection is a very broad subject. In this book I consider convection of the fluids (in particular air and water) at steady state within a horizontal layer, either infinite or in a rectangular box. The upper and lower boundaries are supposed to be either solid or stress free. Even with such limitations, the literature on convection is very large. Below I will mention some of the sources pertaining to the subject of my book.

Henri Claude Benard (1874 –1939), a French physicist, is the "Father" of experimental convection. In his experiments he observed *convection cells*, where liquid arises at the cell's center and drowns at the cell's side, which he found to be vertical.

Lord Rayleigh (1842-1919), a British scientist who made extensive contributions to both theoretical and experimental physics, is the "Father" of theoretical convection. In 1916 he published a paper on convection stability, which up to the present serves as the main instrument in research on stability of various physical processes.

In his paper Rayleigh linearized convection equations and applied to the main unknowns small exponential disturbances. If the disturbances decay with time, then convection is stable, conductive.

If the disturbances grow, convection is unstable.

He introduced what we now call *the Rayleigh number*, Ra, which characterizes the intensity of fluid motion and absorbs five parameters of the fluid. He found its minimal critical value *min*Ra = Ra_{cr} (with the corresponding wave number), which defines the border between stable and unstable convection.

In 1940 Ann Pellew and Richard Southwell published a paper, where they expanded the ideas of Lord Rayleigh. Their method, like his, was based on the linearization of convection equations. With wave number k and using the concept of the Rayleigh number, Pellew and Southwell determined the Neutral Curve, Ra = Ra(k), which divides the plane (k, Ra) into stable and unstable areas.

They found the exact solution for the linearized equation with a wide variety of boundary conditions and cell plainforms. They proved that linear equations can be deduced from the variational principle, from which a simple approximate solution of the same equations can be obtained with high precision.

After the linear theory, convection theory didn't progress much until Landau's 1944 paper, reprinted in Landau & Lifshitz (1987, p.95-98), where the first non-linear theory in fluid dynamics was started, laying the foundation for the whole industry of non-linear theories. One of the fundamental results of that paper was what is known as the *Landau Square Root Law*, which states that above and close to the critical regime, the amplitude of the fluid velocity increases as a square root of the difference between the flow intensity and its critical value.

In 1953, the Russian scientist V.S. Sorokin from Perm University developed a method for complex analysis of the convection equations, which led him to the variational method independent of the form of the container. He showed how the linearized convection equation follows from the variational method.

Introduction

He proved that the linearized equations possess a set of critical Rayleigh numbers - eigenvalues $0 < Ra_1 < Ra_2 < \ldots$ with the corresponding orthogonal eigenfunctions $(V_1, T_1), (V_2, T_2), \ldots$.

Ukhovskii and Yudovich (1963) proved that this set of eigenfunctions composes a complete system. In 1966 and 1967 Yudovich proved the following results for 2D-convection:

- If horizontal velocity is periodical with period $\lambda = 2\pi/k$, then the eigenfunctions of the linearized equations are proportional to $Cos(kx)$ and $Sin(kx)$.

- The velocity flow moving in the horizontal direction through a vertical plane is zero.

- The bifurcation of the solution happens when Ra crosses each eigenvalue Ra_i ($i = 1,2,\ldots$; $Ra_1 = Ra_{cr}$) of the linear solution, but after Ra_2 solutions can be unstable.

- At small $\varepsilon = Ra - Ra_{cr}$ there exist two steady state solutions, they are both asymptotically stable; however, the third equilibrium solution ($V = 0$, $T = az+b$) is unstable.

- Between Ra_1 and Ra_2 both steady state solutions differ by sign only, and both are stable.

- The solution of the classical steady state convection problem exists and it is unique.

- At $Ra = Ra_{cr}$ only a trivial solution exists, which is stable.

  Sorokin (1953) proved it for $Ra < Ra_{cr}$.

In 1965 Schlüter, Lortz, and Busse proved that all 3D-motions in a plane layer are unstable. They also proved that only the 2D-rolls are stable within a certain interval of the wave numbers k. The width of the interval depends on the given Ra. In 1967 Busse found not just the interval, but a closed area ("Balloon Busse") in the plane (k, Ra), where 2D-rolls are stable. Outside of that area the stable steady state does not exist.

One of the reasons for the discrepancy between the experiments and the theories of steady state convection is the different conditions at which they were performed, in particular, usage of the conditions incompatible with the Standard Convection Conditions defined below. Such a definition is necessary because the fields of theoretical and experimental convection overlap only partially.

It is in the common area where the comparison is meaningful.

1. The Standard Convection Conditions

The Standard Convection Conditions describes the conditions for the steady state convection for certain fluids in certain containers.

We define *The Standard Convection Conditions* as those that satisfy the following four points:

1) **Container.**

In theory, the container's simplest geometry is an infinite horizontal layer of the fluid. The depth of the layer, *d,* is the unit of distance.

The liquid layer shouldn't be too deep in order to provide the constancy of liquid characteristics. Usually it is around 1 cm; however, the layer shouldn't be too shallow to prevent the influence of non - homogeneity of the metallic container. In the theory of *Standard Convection Conditions* all the boundaries are ideally homogeneous. In the experiments rectangular, square or cylindrical containers are used with different sizes. In order to be able to compare the experiments in such containers to the theory for the infinite layer it is necessary that the ratio of the longest side L of the container to its depth d, the aspect ratio $\Gamma = L/d$, is larger than 10. This is not an exact number, as it depends on other parameters (like the Prandtl number). Containers with smaller Γ are also used in the experiments and in the theories with the Non-Standard Convection Conditions.

2) **Liquids.** Air, water, liquid helium and silicon oils, are fluids that are used the most in experiments, and will be considered here as incompressible fluids. All their parameters, ρ_0 - density, $\chi = \kappa/(c\rho_0)$ - thermal diffusivity, κ - thermal conductivity, c - heat capacity, α - thermal expansion coefficient, $v = \eta/\rho_0$ - kinematic viscosity, η - dynamic viscosity are considered to be constant. Such liquids are known in literature as *Boussinesq liquids*. Liquids with at least one parameter that is not constant are called *Non - Boussinesq liquids*.

3) **Boundaries.** Upper and lower boundaries can be rigid (for example, copper or plexiglass) or they can be free (as in the experiment with water lying on mercury and covered with oil). Also, the depth d and the temperature difference Θ between lower "hot" and upper "cool" boundaries are assumed to be constant. It is easy in theory, but in the experiments with a thin layer (less than 1 mm) the grain structure of the metallic container prevents homogeneity of the boundaries. Such were the containers described in the review by Bodenschatz *at al.* (2000). Therefore, those experiments are referred to as "Non - Standard Convection".

There are no side boundaries in the infinite layer. In the experimental container, side boundaries are solid. In theory, they could be free. There are different conditions possible on the side boundaries. For solid boundaries, velocity of the liquid satisfies the no - slip condition. Also, solid boundaries are impenetrable. Porous boundaries are not "Standard". Free boundaries are stress free. The temperature on the side boundaries can be a linear function between the lower "hot" and the upper "cold" temperatures. Finally, the most popular case is one with thermal insulated side boundaries. For the Standard Convection Conditions, it is important that the side boundaries are homogeneous, smooth and the conditions on them are stationary.

4) **Convection regime.** In a convecting liquid, there are several dynamic regimes that depend on the given conditions. They differ by the intensity of motion that is described by the dimensionless Rayleigh number $Ra = g\alpha\Theta d^3/(v\chi)$.

Depending on Ra, convection regimes can be different due to the fact that each container has its own increasing sequence of critical Rayleighs: $Ra_1, Ra_2, Ra_3, \dots$. When $Ra < Ra_1$ the liquid is motionless. This is the regime of conduction heat transfer. When $Ra_1 < Ra < Ra_2$, the laminar steady state convection is established. When $Ra_2 < Ra < Ra_3$, the convection becomes time dependent, first as bimodal convection, then as a turbulence. In further increasing Ra, convection develops a strong turbulence.

According to E.L. Koschmieder's (1993) *"moderately supercritical"* regime, the first critical Rayleigh number will be denoted as $Ra_{cr} = Ra_1$.

Introduction

The moderately supercritical regime is such a regime when for a given liquid in the container or in the infinite layer the Rayleigh number lies in the interval $\qquad$ $1.2Ra_{cr} < Ra < 10Ra_{cr}$. $\qquad$ (*)

The left coefficient, $1.2 > 1$, cuts out the area close to Ra_{cr} that, in a sense, divides the stationary and the moving liquid. Near Ra_{cr} various complications can occur due to the increased role of the fluctuations. The right coefficient, 10, cuts out the area close to the next critical Ra_2 to avoid time dependent regime. As we know, many experiments have been performed under the moderately supercritical regime, satisfying this inequality (see Fig.1.1 in Chapter IV). Inequality (*) shows the area where the steady state convection is reached after a certain transitional period. I call this area *Koschmieder's area* because he defined it first.

The inequality (*) is approximate and its numerical coefficients are chosen to cover many experiments. A more precise inequality must include the Prandtl number. In short, the requirement (*) is chosen to make the experiments closer to the idealized theories, apart from the critical regimes. Koschmieder's area is remarkable by the feature that makes all curves $\lambda = \lambda(Ra)$ unique and hysteresis impossible.

As acutely noticed by Ivan Catton (1988): "If the wavenumber path has not proceeded through a bifurcation point prior to reversing the direction of δRa, the original path should be retraced to within the experimental error. On the other hand, if the bifurcation has been crossed, there is no reason to expect retracing of the original path beyond the bifurcation point when the Ra direction is reversed. "

In this text bifurcation points are Ra_1 and Ra_2 . Catton's note means there is no hysteresis between Ra_1 and Ra_2

Moderately supercritical convection, which is considered in this book, is

   a) steady state,
   b) always cellular,
   c) two-dimensional (2D).

Cells periodically repeat themselves and manifest themselves in the form of straight rolls, characterized by the *wave number k*. One cell consists of two adjacent rolls with the fluid rising between the roll axes and sinking along the vertical boundary of a cell. Fluid motion in this regime is two-dimensional, so I choose x as the horizontal axis and z as the vertical axis. The diameter of a cell is $\lambda = 2\pi/k$.

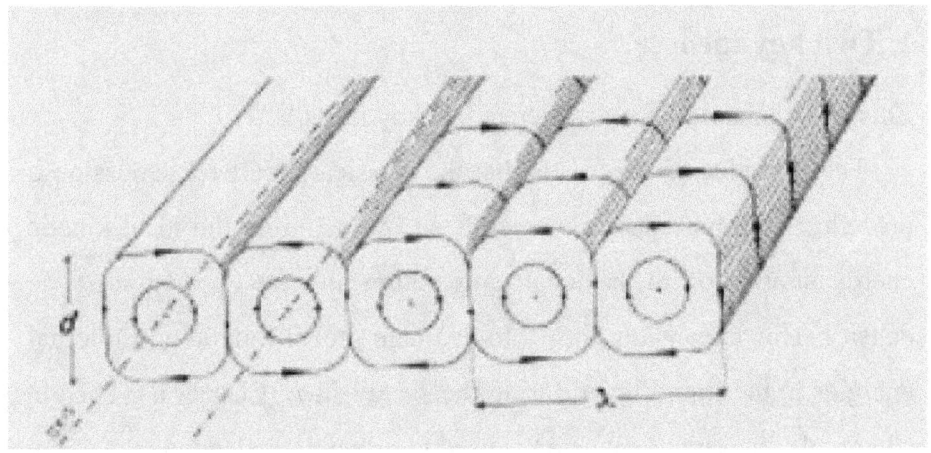

Fig 1. Convection rolls.

The cell 2D-volume is

$$\Omega = \{ -\pi/k \leq x \leq \pi/k, \ 0 \leq z \leq 1 \} \qquad (**)$$

Our goal is to find how the cell size λ depends on the temperature difference, or to find the function $\lambda = \lambda(\mathrm{Ra})$.

An infinite layer along with constant horizontal boundary conditions makes all cells identical, so, we need to study just one cell. We know what kind of conditions must be accepted for the vertical lateral boundary between cells. We accept the usual assumption, that at steady state:

a) The cell boundary is a vertical plane (which is not true, but almost true) – (Bergé, 1975) and

b) During the steady state there is no exchange of kinetic or internal energy between two adjacent cells. In other words, all the cells are the same not only by geometry, but by physics as well.

2. Two key ideas

This book is based on two independent key ideas.

The first idea (Chapter I) is that the classical Oberbeck-Boussinesq procedure (OB-procedure) is generalized by including the local kinetic energy in addition to the local temperature into the expansion of the density. This expansion allows to evaluate global kinetic and internal energies in the liquid layer during the steady state (Chapter II). Then in Chapter II, the same expansion is used to prove the independence of the steady state on any initial conditions. In chapter III the results of numeric experiments are presented that confirm independence of the steady state convection from initial conditions.

The second idea is the Principle of Maximum Entropy Production. Chapter IV shows how we can predict the cell-size dependence on heating temperature by using this principle.

I. EXPANSION OF DENSITY AND BASIC EQUATIONS

The mathematical model of convection includes the system of equations for momentum and for heat transfer of the moving fluid. Besides the main dynamic variables like velocity, temperature and pressure, there are many physical parameters characterizing various fluid properties.

Convection equations are so complex that they are sometimes characterized by the sentence: "No matter how much you simplify them, they still remain very complex." So, it is natural that the first step in their simplification is to make all fluid parameters constant, such as specific heat capacity, kinematic viscosity and so on. For the fluid layer of small thickness, we can make constant all parameters, except one – fluid density. Fluids with constant density are called *incompressible*. Otherwise – *compressible*. Equations for convection in compressible fluids are so complex that they allow only a numerical solution. That is why most research is done for incompressible fluids.

However, the simple truth about incompressible fluids is that in them **convection is impossible**. Indeed, convective motion starts when some part of liquid volume is lighter than another. Light parts rise due to buoyancy while heavier parts drown. In other words, it is necessary for convection that the density is not constant. Non-constant density makes convection equations extremely complex.

1. **The OB-procedure**

To simplify the situation, the German physicist Anton Oberbeck in 1879 and the French mathematician Joseph Boussinesq in 1903 independently introduced what we now call the Oberbeck-Boussinesq procedure (OB-procedure). They began with the general hydrodynamic Navier-Stokes equation for compressible liquid in a gravitational field:

$$\rho(\ \partial V/\partial t + (V\nabla)V\ ) = -\nabla P + \eta\Delta V + \rho g\ . \tag{1.1}$$

Then they expanded the variable density ρ into a power series of small temperature deviations T' from the constant mean temperature T_0 corresponding to the same fluid with constant density ρ_0 :

$$\rho = \rho_0(1 - \alpha T') \tag{1.2}$$

where ρ_0 and α are constant density and thermal expansion.

Because ρ-dependence on pressure P is insignificant qualitatively and quantitatively, P is omitted in the expansion (1.2).

Finally, they replaced ρ by ρ_0 everywhere in (1.1) except the driving force ρg where density ρ was replaced by its expansion (1.2). They replaced ρ by ρ_0 in the heat equation and in the continuity equation. This is the OB-procedure. As a result, we obtain the OB-equations:

$$\rho_0(\ \partial V/\partial t + (V\nabla)V\ ) = -\nabla P + \eta\Delta V + g\rho_0(1 - \alpha T')\ , \tag{1.3}$$

$$c\rho_0(\partial T'/\partial t + V\nabla T') = \kappa\Delta T', \tag{1.4}$$

$$\mathrm{div}V = 0. \tag{1.5}$$

T' is the temperature deviation from the constant temperature T_0 .

Now all physical parameters, including ρ_0, are constant, yet the driving force depends on the variable temperature T' making convection possible.

Chapter I. **Expansion of density**

Note that the OB-procedure is not a formal logical derivation; it is a recipe (an equation with ρ_0 on the left, and with (1.2) on the right), and that's why it is called a procedure. Navier-Stokes equation (1.1) and OB-equation (1.3) differ not only mathematically, but physically as well. The former has a buoyancy ρg as a driving force of pure potential energy, while the latter has thermodynamical T' acting on a gravitational force g and driving convection.

The OB-equation (1.3) is a physical model for the Navier-Stokes equation (1.1) with a different physical mechanism. The model with a different mechanism for incompressible fluid supposes to approximate the original mechanism for compressible fluid.

In fact, the model became so good that for a hundred years it served well for many various applications.

2. The extended OB-procedure

However, there are some questions that cannot be answered by the OB-equations. One such important question is about the influence of the initial conditions on the steady state convection. There is a school of thought that says the so-called controlled initial conditions can create a steady state of any form with arbitrary number of rolls (Chen and Whitehead, 1968, also Busse and Whitehead, 1971). In his book Koschmieder (1993) criticized controlled initial conditions from the experimental point of view. In Chapter II of this book I will provide the proof that steady state convection does not depend on initial conditions at all. This proof is based on a procedure similar to the OB-procedure, but with a new expansion of density.

Strictly speaking, density change occurs not only due to the temperature and the pressure, but also due to the presence of the velocity V. Quantitatively the influence of V onto ρ is small. Nevertheless, it will play a decisive role in finding the convection global kinetic and internal energies. (See Note 2 after equation (4.4)).

The expansion of density ρ into a power series of velocity cannot begin with the term proportional to V, because ρ as a scalar cannot be proportional to the vector V. Even such a scalar product as aV with vector a cannot be part of the Taylor expansion, because $a = \partial\rho/\partial V$ depends on V, while the only vector-parameter in the convection equations is the vector of the gravitational acceleration g, which is independent of V. Thus, the expansion of ρ into power series of V must begin with a constant and continue with the term V^2. Combining temperature and velocity in one expansion, we obtain

$$\rho = \rho_0(1 - \alpha T' + \beta V^2/2!).$$

Note that each term inside the parenthesis must be dimensionless because density ρ_0, outside the parenthesis, has absorbed all the necessary dimensions. To make βV^2 dimensionless β must have the dimension of $1/V^2$. In fluids, the only non-compound physical value with dimension of V is the speed of sound, u. Therefore, I rewrite our density as

$$\rho = \rho_0(1 - \alpha T' + V^2/(2u^2)) \tag{2.1}$$

where u is the speed of sound in the given fluid. Similar to this expansion is an expansion found in Landau & Lifshitz (1986, §10, p.21), where they write the change in density as $\Delta\rho_0 \sim \rho_0 V^2/c^2$ with V as the speed of fluid and c as the speed of sound.

Equation (2.1) connects thermodynamic variables ρ_0 and T' with mechanical energy $V^2/2$. This means that density is expanded in the linear function of internal energy and kinetic energy. We have a similar connection in the *Bernoulli law* $V^2/2 + P/\rho_0 + gz = constant.$

The last term $V^2/(2u^2)$ in (2.1) is so small compared to others that it will not play a role in the numerical calculations of steady state, but it will play the main role in finding the convection global kinetic and internal energies. A procedure with a density such as (2.1) leads to the equation

$$\partial V/\partial t = - (V\nabla)V - \nabla P/\rho_0 + v\Delta V + g(- \alpha T + V^2/(2u^2)) \qquad (2.2)$$

It is this equation along with the rest of OB-equations and boundary conditions that allow to prove (in Chapter II) the independence of steady state from any initial conditions. Furthermore, my numerical calculations (in Chapter III) as well as earlier calculations by Theodore D. Foster (1969, see graph at the end of the Chapter III) both confirm the idea of independence.

3. Boundary conditions

There are two types of boundary conditions corresponding to two types of boundaries.

Solid boundaries are impermeable therefore velocity perpendicular to the boundary vanishes at the boundary. Velocity parallel to the boundary is subject to no-slip condition. Liquid molecules are attracted to the solid boundary by electrical forces and cannot move.

For a rigid boundary no-slip conditions can be written as:

$$V = 0, \qquad z = 0, d; \tag{3.1}$$

Since this condition must be satisfied for all x on the surface, it follows from the equation of continuity, $\partial_z V_z + \partial_x V_x = 0$, that on a rigid surface

$$\partial_z V_z = 0, \qquad z = 0, d. \tag{3.2}$$

Free boundaries are not a plane, but very close to it. As it is indicated by Drazin and Reid (1984) "… a free surface behaves as a rigid surface with tangential slip but without any tangential stress." This implies the vanishing of V_z:

$$V_z = 0, \qquad z = 0, d \tag{3.3}$$

and the vanishing of tangent *(to the plane horizontal surface)* stress. This was assumed by Rayleigh (1916) and supported by Davis and Segel (1968). Having a vanishing tangent *(to the plane horizontal surface)* stress $\partial_x V_z + \partial_z V_x = 0$, and having $V_z = 0$ for all x, we can write $\partial_z V_x = 0$ or substituting it into $\partial_z (div V) = \partial_{zz} V_z + \partial_{zx} V_x = 0$ we have free boundary conditions: $\qquad \partial_{zz} V_z = 0, \qquad z = 0, d. \tag{3.4}$

4. The convection model

In this chapter the extended OB-procedure will be used with approximation (2.2) instead of (2.1). I will write the system of convection equations for the "separated" (from conduction) temperature and pressure.

If $T_{cnd}(z) = T_h - z(T_h - T_c)/d$ and P_{cnd} satisfying $\nabla P_{cnd} = \rho_0 g(1 - \alpha T_{cnd})$, then the corresponding deviations - convective temperature T and pressure P are counted from $T_{cnd}(z)$ and P_{cnd}.

Chapter I. **Expansion of density**

Thus, the system (convection model) we will study is:

$$\partial V / \partial t + (V\nabla)V = -\nabla P / \rho_0 + v\Delta V + g(-\alpha T + V^2/(2u^2)) , \quad (4.1)$$

$$\partial T / \partial t + V\nabla T = \chi \Delta T + aV_z , \qquad a = (T_h - T_c)/d . \quad (4.2)$$

$$\mathrm{div} V = 0. \quad (4.3)$$

$$T = P = V_z = \partial_z V_z = 0 \quad \text{at } z = 0 \text{ and } z = d. \quad (4.4)$$

The condition $P = 0$ is a consequence of three facts.

First, we excluded hydrostatic pressure P_{cnd}.

Second, since our fluid is separated from its environment by the solid boundaries, such an environment cannot put pressure onto the fluid. Third, inside the fluid layer the temperature can influence the pressure, but since the boundary temperature is zero, so is the boundary pressure.

Note 1. In the equation of momentum (4.1) the coefficient at V^2 is so small, that the whole term $V^2/(2u^2)$ can be discarded within a numerical solution.

Note 2. However, $1/(2u^2)$ plays a different role in local and global equations. To illustrate the situation by analogy, consider the comparison of two functions f and φ : $f = 100\mathrm{Sin}(x)$, $\varphi = \varepsilon \mathrm{Sin}^2(x)$ on the interval $\Omega = (0, 2\pi)$. With a very small ε the function φ is close to zero, while f is not so small $100\mathrm{Sin}(x)$. However, when we compare the integrals of those two functions on Ω, then the integrals are 0 and $\varepsilon\pi$ correspondingly. A similar situation can be observed, when we compare the two last terms in the momentum equation (4.1): when we use that equation for velocity calculations, that is locally, we can drop the term $V^2/(2u^2)$, and that is what happens in the classical convection theory. However, when we integrate the momentum equation at the steady state, we arrive at the global internal and global kinetic energies of the same order.

Note 3. There is a great numerical difference between coefficients in (4.1), because they belong to variables of a different nature - temperature and squared velocity. When we reformulate

$$\alpha T - V^2/(2u^2) \qquad (4.5)$$

in terms of similar physical variables like internal and kinetic energies, their coefficients will be of the same order. Indeed, local internal and kinetic energies are $e_{in} = \rho_0 cT$, $e_{kin} = \rho_0 V^2/2$, from where

$$T = e_{in}/\rho_0 c, \quad V^2/2 = e_{kin}/\rho_0. \qquad (4.6)$$

Substituting this into (4.5), one obtains:

$$(\alpha/\rho_0 c)e_{in} - e_{kin}/(\rho_0 u^2) \qquad (4.7)$$

In (4.5) there is a great difference between the coefficients of temperature and squared velocity, but in (4.7) there is no difference between the order of coefficients of internal and kinetic energies.

For example, in water, coefficients of T (measured in $°K$) and V^2 (measured in cm^2/s^2) in (4.5) are correspondingly of order 10^{-4} and 10^{-10}, while in (4.7) both coefficients (measured per unit volume) of e_{in}, e_{kin} are of the order correspondingly 10^{-11} and 10^{-10} ($\alpha \sim 10^{-4}$, $c \sim 10^{7}$, $u \sim 10^{5}$.)

Note 4. It is necessary to stress the following: neither the OB-expansion $\rho = \rho_0(1 - \alpha T')$, nor my expansion (2.1) is the equation of state for the incompressible liquid. Its true equation of state is:

$$\rho_0 = constant. \qquad (4.8)$$

Indeed, substituting (2.1) or (2.2) into the compressible continuity equation $\partial\rho/\partial t + div(\rho V) = 0$, we will get an equation, which contradicts the heat transfer equation (4.2).

5. Convective kinetic and internal energies

To get the connection, it is sufficient to use the momentum equation at the steady state ($\partial V/\partial t = 0$):

$$(V\nabla)V = -\nabla P/\rho_0 + v\Delta V + g(-\alpha T + V^2/(2u^2)), \tag{5.1}$$

Integrating it over the whole volume occupied by the liquid (see Addendum), and retaining the non-zero terms, we get the equation ($d\Omega$ is an element of the volume):

$$\int(\alpha T - V^2/(2u^2))d\Omega = 0 \tag{5.2}$$

or using (4.6) with $E_{in} = \int e_{in}d\Omega$, $E_{kin} = \int e_{kin}d\Omega$, we obtain

$$(\alpha/\rho_0 c)E_{in} - E_{kin}/(\rho_0 u^2) = 0. \tag{5.3}$$

From this we get:

$$E_{kin}/E_{in} = \alpha u^2/c. \tag{5.4}$$

The right-hand side of (5.4) is a constant, which means that *during the steady state, the global internal convective energy and the kinetic energy are proportional.*

The left part of (5.4) is a dimensionless number, the right part is also dimensionless. Thus, *at steady state, the ratio of kinetic and internal energies depends neither on the boundary temperatures nor on density, viscosity or thermal conductivity.*

For example, water parameters at 20°C are $\alpha = 2.07*10^{-4}$, $c = 4.18*10^7$, $u = 1.48*10^5$, which leads to

$$\alpha u^2/c = 0.108 \quad \text{or} \quad E_{kin} = 0.108 E_{in}. \tag{5.5}$$

It is interesting to get an equality similar to (5.4) in terms of T and V^2.

For this one can use (5.2): $\int(\alpha T - V^2/(2u^2))d\Omega = 0$, from which

$$\int T \, d\Omega / \int V^2 d\Omega = 1/(2u^2\alpha). \tag{5.6}$$

Taking for u and α the same numbers as above, we get

$$1/(2u^2\alpha) = 1.1*10^{-7}. \tag{5.7}$$

This example tells us that comparison of T and V^2 hugely depends on the units of T. When T is expressed in terms of temperature then it is overwhelmingly less than V^2 (10^7 less). However, when both T and V^2 are expressed in the same units – units of energy – then they are almost of the same order (5.5).

6. The paradox of density

Consider a container filled with a compressible liquid with density

$$\rho = \rho_0(1 - \alpha T' + V^2/(2u^2)), \text{ which I rewrite as,}$$
$$\rho = \rho_0 - \rho_0(\alpha T' - V^2/(2u^2)), \tag{6.1}$$

where $\qquad \rho_0 = (\int\rho d\Omega)/\Omega$ is the average constant density.

Integrating (6.1) over the whole container's volume, we obtain:

$$M = M - \rho_0\int(\alpha T' - V^2/(2u^2)) d\Omega, \tag{6.2}$$

where M is the total mass of the liquid within volume Ω.

The last equality can be rewritten as

$$\int(\alpha T' - V^2/(2u^2)) d\Omega = 0, \tag{6.3}$$

which is valid **at any time**. This contradicts (5.2), which is valid **for the steady state only**. And that is the Paradox of density.

Resolution of the Paradox of density comes from the fact that the basis for the Paradox, equation (6.1), is not an equation of state and therefore is not a part of the model for compressible or incompressible liquids. Equation (6.1) is part of the extended OB-procedure.

In other words, equation (6.1) is an instrument used for transitioning from the model of compressible liquid to the model of incompressible liquid.

I can imagine an analogue situation, when I use an ax in the forest to build a house. The ax is neither part of the forest, nor of the house.

It's just an instrument useful for building the house. The same is with equation (6.1).

II. INDEPENDENCE OF STEADY STATE CONVECTION FROM INITIAL CONDITIONS

There is no consensus in the scientific community about whether or not steady state convection depends on initial conditions. In his book Koschmieder (1993) provided a review of this matter and analyzed some important experiments. To summarize, those experiments that did show dependence of steady state convection on initial conditions didn't last long enough to permit the fluid to adjust itself to the independent state. Another reason for the discrepancy between the experiments and the theories is the different conditions at which they were performed, in particular, usage of the conditions incompatible with the Standard Convection Conditions defined in the Introduction to this book.

This chapter heavily leans on the previous chapter, where concepts of expanded density and the extended model of convection were developed. In particular, the approximation (I.2.2) will be used instead of the traditional Oberbeck-Boussinesq approximation.

1. The main equations

I will write the system of convection equations for the "separated" (from conduction) temperature and pressure.

If $T_{cnd}(z) = T_h - z(T_h - T_c)/d$ and pressure P_{cnd} satisfies $\nabla P_{cnd} = \rho_0 g(1 - \alpha T_{cnd})$, then the corresponding deviations - convective temperature T and pressure P are counted from $T_{cnd}(z)$ and P_{cnd}.

Thus, the system from Chapter I that we study is rewritten:

$$\partial V/\partial t = -(V\nabla)V - \nabla P/\rho_0 + v\Delta V + g(-\alpha T + V^2/(2u^2)), \qquad (1.1)$$

$$\partial T/\partial t = -V\nabla T + \chi\Delta T + aV_z, \qquad a = (T_h - T_c)/d. \qquad (1.2)$$

$$\mathrm{div}V = 0. \qquad (1.3)$$

$$T = P = V_z = \partial_z V_z = 0 \qquad \text{at } z = 0 \text{ and } z = d. \qquad (1.4)$$

Here u is the speed of sound in the given fluid, d is the depth of the liquid layer, and all the rest of the constant coefficients have standard meaning.

2. Fundamental theorems of steady state

I will list below well-known Fundamental Theorems plus a new one. They are called "Fundamental" because each of them gives necessary and sufficient conditions for the steady state convection in terms of kinetic energy, entropy, internal energy, and dissipation.

Each Theorem is presented as an integral equality with temperature and velocity in dimensionless form. All four are independent of pressure.

1. The first one is a necessary and sufficient condition of the conservation of kinetic energy (Chandrasekhar, 1981).

Multiplying the Momentum equation (1.1) by V, integrating the result over the cell volume, and using the conservation of mass (1.3) and the boundary conditions (1.4), we get the necessary and sufficient condition of conservation of kinetic energy (see Addendum):

$$\partial_t \int (V^2/2)d\Omega = \int (-(\partial_i V_j + \partial_j V_i)^2/2 + RaV_z T)d\Omega = 0, \qquad (2.1)$$

where summation is assumed for expressions with repeated indices $(i=x,z;\ j=x,z)$. This equation is called ***Chandrasekhar's Theorem of Balance*** or ***The First Theorem of Balance***.

2. Similarly, multiplying the internal energy equation (1.2) by T, integrating the result over the cell volume, and using the conservation of mass (1.3) and the boundary conditions (1.4), we get the necessary and sufficient condition of conservation of internal energy

$$\text{Pr}\cdot\partial_t \int (T^2/2)d\Omega = \int (V_z T - (\boldsymbol{\nabla} T)^2)\, d\Omega = 0. \qquad (2.2)$$

We will call this equation ***The Second Theorem of Balance***.

It establishes the conservation of energy indirectly, through conservation of the square of temperature.

3. Direct conservation comes from the equation of energy (1.2). Integrating it, one obtains

$$0 = \text{Pr}\cdot\partial_t \int T d\Omega = \int \Delta T d\Omega + \int V_z\, d\Omega. \qquad (2.3)$$

The last integral in (2.3) is proportional to the vertical flux of liquid leaving volume Ω. That flux is zero. The first integral on the right side can be reduced to the surface integral:

$$0 = \text{Pr} \cdot \partial_t \int T d\Omega = \int \Delta T d\Omega = \int \nabla T (z{=}1) dx - \int \nabla T\ (z{=}0) dx\ , \qquad (2.4)$$

where the integral on the left side is taken over the cell volume $\Omega = \{-\pi/k \le x \le \pi/k,\ 0 \le z \le 1\}$, but the two integrals on the right side are taken over x from $x = -\lambda/2$ to $x = +\lambda/2$, where $\lambda/2$ is the roll diameter and λ is the cell diameter. The heat flux incoming to the volume Ω is equal to the heat flux outgoing from Ω. So, the equality of incoming and outgoing fluxes, (2.4), is *the necessary and sufficient condition for the conservation of internal energy.*

And this is *the Third Theorem of Balance.*

4. Combining the 1st and 2nd theorems, one can obtain *the Fourth Theorem of Balance:*

$$\int [-(\partial_i V_k + \partial_j V_i)^2/2 + \text{Ra}(\nabla T)^2\ ] d\Omega = 0. \qquad (2.5)$$

The first term under the integral here is the Dissipative function. The second term is the Entropy production. In different situations one or the other of these Theorems will be useful.

3. Two Nusselt numbers

During the free convection, in particular during the steady state convection in a horizontal layer of liquid, the heat transfers from the lower hot plane to the upper cold plane and further to the environment over that upper plane. Such a heat transfer occurs in two ways simultaneously. One way is the conduction, another is convection.

Let Q_{cond} be a heat flux transferred per unit area and per unit time by the thermal conductivity of the resting fluid alone, and Q_{conv} is a heat flux transferred additionally by the convective motion of the fluid.

In convection theory and in experiments an important role is played by the *Nusselt number,* which is defined as

$$Nu = (Q_{cond} + Q_{conv})/ Q_{cond}. \qquad (3.1)$$

However, in an experiment and in a theory, there are two different definitions, which I will call the First and the Second. The First is based on heat flux through the surface. The Second is based on the heat flux through the volume. The First are mostly used by experimenters from those earlier experiments described by Chandrasekhar (1981) through later ones such as the experiments by Koschmieder & Pallas (1973). The theorists are using exclusively the Second definition.

The purpose of this section is to prove that both definitions are equivalent.

The First Nusselt number

In his well-known book S. Chandrasekhar (1981, §18, page 61) writes: "Let $\mathcal{L}$ denote the flux of heat: it measures the quantity of **heat emerging from the surface**, per unit area and per unit time. If the flow of heat across the fluid layer is entirely by conduction, then clearly

$$\mathcal{L} = \chi \Delta T/d, \qquad (3.2)$$

where χ is the coefficient of heat conduction."

Here ΔT is the temperature difference between lower and upper plane.

Chandrasekhar's χ is κ in modern notations. Modern $\chi = \kappa/(\rho c)$, which is called *coefficient of thermodiffusion*. Note the key words "heat emerging from the surface".

Two pages further Chandrasekhar writes:

" $\mathscr{L}$ expressed in the unit $\chi \Delta T/d$ is often called the **Nusselt number**:

$$\text{Nu} = \mathscr{L}/(\chi \Delta T/d). \text{ "}$$

(3.3)

Here $\mathscr{L}$ is the total flux, including conduction and convection parts, not just (3.2). Again, this definition of Nu operates with the flux of the heat "emerging from the surface".

Similar to that, but more general definition can be found in Landau & Lifshitz (1986, §53) :

"The heat transfer between solid bodies and the fluids is usually characterized by the *heat transfer coefficient α*, defined by

$$\alpha = q/(T_1 - T_0),$$

(3.4)

where q is the **heat flux density through the surface** and $T_1 - T_0$ is a characteristic temperature difference between the solid body and the fluid. If the temperature distribution in the fluid is known, the heat transfer coefficient is easily found by calculating the heat flux density $q = -\kappa \partial T/\partial n$ at the boundary of the fluid (the derivative being taken along the normal to the surface).

The heat transfer coefficient is not dimensionless quantity.

A dimensionless quantity which characterizes the heat transfer is the *Nusselt number:*

$$\text{Nu} = l\alpha/\kappa. \text{ "}$$

(3.5)

Here l is characteristic length, κ is a coefficient of thermoconductivity. Note the key words "q is the heat flux density through the surface".

Thus, in both cases – Chandrasekhar's and Landau and Lifshitz' Nusselt number is defined by **the flux through the surface** and in accordance to (3.1).

The Second Nusselt number

In most of the literature on the theoretical convection the Nusselt number is defined as a heat flux transferred through the whole volume of liquid layer, that is

$$\text{Nu} = 1 + \int V_z T dx dz, \tag{3.6}$$

where integration is over the box $\{0 \leq x \leq \Gamma, \ 0 \leq z \leq 1\}$.

In (3.6) V_z and T are 2D-functions satisfying the following conditions:

$$\partial V/\partial t + (V\nabla)V/\text{Pr} = -\nabla P + \Delta V + \gamma \text{Ra}T, \tag{3.7}$$

$$\text{Pr}\cdot\partial T/\partial t + V\nabla T = \Delta T, \tag{3.8}$$

$$\text{div}V = 0. \tag{3.9}$$

$$T = 1 \text{ at } z = 0; \ \ T = 0 \text{ at } z = 1; \ \ V_z = \partial_z V_z = 0 \ \text{ at } z = 0 \text{ and } z = 1. \tag{3.10}$$

$$\partial_x T = V_x = \partial_x V_x = 0 \ \text{ at } x = 0 \text{ and } x = \Gamma. \tag{3.11}$$

Here $\text{Ra} = ga\Theta d^3/(v\chi)$ - Rayleigh number, $\text{Pr} = v/\chi$ - Prandtl number.

For the cell $\Omega = \{ -\pi/k \leq x \leq \pi/k, \ 0 \leq z \leq 1\}$, the conditions are:

$$V_x(x,z) = V_{xx}(x,z) = \partial T(x,z)/\partial x = 0, \qquad x = -\pi/k \text{ or } \pi/k. \tag{3.12}$$

Here k is a wave number $k = 2\pi/\lambda$ and λ is a cell diameter.

We also have symmetry conditions:

$$V_z(-\pi/k \leq x \leq 0, \ z) = V_z(0 \leq x \leq \pi/k, \ z), \ T(-\pi/k \leq x \leq 0, \ z) = T(0 \leq x \leq \pi/k, \ z). \tag{3.13}$$

Our constant density ρ_0 was defined at an average constant temperature T_{av} and the corresponding hydrostatic pressure satisfying $\nabla P_0 = \rho_0 \mathbf{g}$.

In our main equations above the temperature T is a deviation from T_{av} and pressure P is a deviation from P_0.

Dimensional (designated with stars) time, pressure, velocity, and temperature connected to dimensionless values are:

$$t^* = t(d^2/v), \quad P^* = P(\rho_0 v \chi / d^2), \quad V^* = V(\chi/d), \quad T^* = T\Theta,$$

$$(x^*, y^*, z^*) = (x, y, z)d \qquad 3.14)$$

The form (3.6) is due to the chosen dimensionalization (3.14). Because of that particular choice, the Rayleigh number didn't appear explicitly in the energy equation (3.8) nor in the Nusselt number (3.6).

With different choice of dimensionalization the Nusselt number could involve the Rayleigh number and the Prandtl number. In any case the Nusselt number depends on the functions V and T that in turn depends on Ra and Pr.

The equivalence of two Nusselt numbers

Chandrasekhar (1981, Appendix I) did not pose an equivalence problem, but he developed all the necessary techniques for the proof of the problem in the 3D case. I will use his technique for the 2D case in dimensionless form with different notation and with more details. The generalization for the 3D case is trivial.

We will operate assuming the steady state regime of convection. This means all the time derivatives are zero, convection cells are adjacent rolls, velocity and temperature are periodic functions of x with period λ. Convection is assumed to be in infinite horizontal layer of the fluid or in the 2D-container with the fluid with the boundary conditions specified above.

Chapter II. **Independence of steady state**

As we have established, both Chandrasekhar and Landau-Lifshitz defined Nusselt number as the total heat flux leaving the surface of container and normalized to the heat flux for the resting fluid in the same container.

Let's begin with rewriting their definition in terms of T, defined in (3.8). Dimensionless heat flux leaving upper boundary and integrated over whole length $0 \le x \le \Gamma$ is:

$$Q = - \int [(\partial T/\partial z)_{z=1}]dx. \tag{3.15}$$

The minus sign is here because the flux direction is opposite to the direction of increasing temperature. This flux includes a convective part as well as a conductive part because we took T from the above equations where both parts were presented inseparable.

As conductive temperature is simply $T_{cond} = 1 - z$, the corresponding conductive flux $Q_{cond} = -1$. Therefore, the Nusselt number is (we supply index 1 meaning the Nusselt first definition):

$$Nu_1 = |Q/(-1)| = |Q|. \tag{3.16}$$

Our purpose is to prove that this definition is the same as (3.6).

To do that we need just one equation (3.8) and boundary conditions (3.10), (3.11):

$$\boldsymbol{V}\boldsymbol{\nabla}T = \Delta T, \tag{3.17}$$

$$T = 1 \text{ at } z = 0; \quad T = 0 \quad \text{at } z = 1, \tag{3.18}$$

$$\partial_x T = 0 \quad \text{at } x = 0 \text{ and } x = \Gamma. \tag{3.19}$$

The heat equation (3.17) is written for the steady state, so time derivative is zero.

Let's rewrite (3.17) in the form more suitable for our treatment:

$$\boldsymbol{V}\boldsymbol{\nabla}T = \text{div}(\boldsymbol{V}T) - T\text{div}\boldsymbol{V} = \partial_x(V_x T) + \partial_z(V_z T) = \partial_{xx}T + \partial_{zz}T. \tag{3.20}$$

During the steady state identical cells completely fill the container or infinite horizontal layer of the fluid. Therefore, it is usually assumed that V and T are periodic functions of x and can be presented as

$$[V] = 0, \quad [T] = T_0(z), \quad T(x,z) = T_0(z) + \theta(x,z), \quad [\theta(x,z)] = 0, \qquad (3.21)$$

where brackets notate integration over x either from 0 to Γ or over horizontal diameter of the cell.

After integrating (3.21) by x it is obvious that

$$T_0(0) = 1, \quad T_0(1) = 0, \quad \theta(x,0) = \theta(x,1) = 0. \qquad (3.22)$$

Now I will follow the key idea of Chandrasekhar (1981, Appendix I), and I will integrate (3.20) over x from 0 to Γ using boundary conditions (3.19):

$$\partial_z[V_z\theta] = \partial_{zz}[T_0], \qquad (3.23)$$

where brackets notate integration over x.

Equation (3.23) is an ordinary differential equation of the second order. To get a Nusselt number I need the flux $\partial_z[T]$. So, I will solve (3.23) in order to use boundary conditions (3.18), (3.22) and then I will differentiate my solution by z to obtain a flux and Nusselt number.

The general solution of (3.23) is (C1 and C2 are arbitrary constants):

$$[T_0](z) = \int_0^z [V_z\theta]dz + zC1 + C2. \qquad (3.24)$$

Using boundary conditions (3.18) we find

$$[T_0](z) = \int_0^z [V_z\theta]dz - z\langle V_z\theta\rangle - z + 1, \qquad (3.25)$$

where angular brackets denote integration over whole volume of container (or cell):

$$\langle V_z\theta\rangle = \int_0^\Gamma \int_0^1 (V_z\theta)dx\,dz, \qquad (3.26)$$

To obtain the Nusselt number we need the heat flux to leave the upper boundary, so we differentiate (3.25):

$$\partial_z[T_0] = [V_z\theta] - \langle V_z\theta \rangle - 1, \tag{3.27}$$

Taking this expression at $z = 1$ where $V_z = \theta = 0$, we have due to (3.22):

$$\partial_z[T_0]_{z=1} = -\langle V_z\theta \rangle - 1 \tag{3.28}$$

or $\qquad \left| \partial_z[T_0]_{z=1} \right| = \langle V_z\theta \rangle + 1 . \tag{3.29}$

The left part of (3.28) is the heat flux exiting the upper boundary $z = 1$, so direction of flux is the opposite to the sign of the derivative $\partial_z[T_0]$, which is negative because the right side of (3.28) is negative, since $\langle V_z\theta \rangle > 0$ due to the second Fundamental Theorem (2.2).

As the left part of (3.29) includes conductive and convective flux and as $\langle V_z\theta \rangle$ is just convective flux while $|-1|$ represents conductive flux, we have according to definitions (3.5), (3.6):

$$Nu_1 = Nu_2 . \tag{3.30}$$

The statement is proven.

4. The independence of steady state convection from initial conditions

We begin this section without assuming the steady state.
First, let's rewrite our system (1.1)-(1.4):

$$\partial V/\partial t = -(V\boldsymbol{\nabla})V - \boldsymbol{\nabla}P/\rho_0 + v\Delta V + g(-\alpha T + V^2/(2u^2)) , \tag{4.1}$$

$$\partial T/\partial t = -V\boldsymbol{\nabla}T + \chi\Delta T + aV_z , \qquad a = (T_h - T_c)/d . \tag{4.2}$$

$$\mathrm{div}V = 0. \tag{4.3}$$

$$T = P = V_z = \partial_z V_z = 0 \qquad \text{at } z = 0 \text{ and } z = d. \tag{4.4}$$

Now I will take equation (5.2) from Chapter I:

$$\int(\alpha T - V^2/(2u^2))\ d\Omega = 0. \tag{4.5}$$

and after differentiating it by t substitute equations of internal and kinetic energy in the result. So, I start with

$$\alpha\int\partial_t T d\Omega - (1/u^2)\int\partial_t V^2/2 d\Omega = 0. \tag{4.5a}$$

To obtain internal and kinetic energy equations I'll multiply (4.1) by V/u^2 and (4.2) by α and integrate the results over volume Ω. Writing nonzero terms (see details of integration in Addendum) we have:

$$\partial_t\int(V^2/(2u^2))d\Omega = (1/u^2)\int(-v(\partial_i V_j + \partial_j V_i)^2/2 + \alpha g V_z T)d\Omega, \tag{4.6}$$

$$\alpha\partial T/\partial t = \alpha\chi\int[(\partial T/\partial z)_{z=1}]dx - \alpha\chi\int[(\partial T/\partial z)_{z=0}]dx =$$
$$= \alpha Q_c - \alpha Q_h = \alpha\Delta Q, \tag{4.7}$$

Substituting (4.6) and (4.7) into (4.5a) we have:

$$\partial_t\int(\alpha T - V^2/(2u^2))d\Omega = \alpha\Delta Q - (1/u^2)\int(-v(\partial_i V_j + \partial_j V_i)^2/2 + \alpha g V_z T)d\Omega, \tag{4.8}$$

This is a rather cumbersome expression.

Let's introduce simplifying notations:

$$E(t) = \int(\alpha T - V^2/(2u^2))d\Omega, \quad D(t) = \int(\partial_i V_j + \partial_j V_i)^2/2 d\Omega, \quad B(t) = \int V_z T\ d\Omega. \tag{4.9}$$

With these notations (4.8) can be rewritten as

$$\partial E/\partial t = \alpha\Delta Q - (1/u^2)(-vD + B) = N(t), \tag{4.10}$$

where new function $N(t)$ of time t was introduced for further simplification:
$$\partial E/\partial t = N(t). \tag{4.11}$$

Now I'll integrate it over time t in order to get a solution explicitly including initial conditions:

$$E(t) = E_0 + \int_0^t N(t')dt'. \tag{4.12}$$

Here the constant E_0 is the initial condition, a combination of initial conditions for temperature and velocity.

Chapter II. **Independence of steady state**

During the evolution from initial moment through the steady state, it is possible to distinguish two periods: the first is the transitional period starting from $t = 0$ to the beginning of the steady state $t = t_s$, the second is a steady state $t \geq t_s$ coming right after the transition. In accordance with this, one can rewrite E as a sum:

$$E(t) = E_1(t) + E_2(t) \tag{4.13}$$

with $\quad E_1(t) = E_0 + \int_0^{t_s} N(t)dt; \quad E_2(t) = \int_{t_s}^t N(t)dt . \tag{4.14}$

Since the E_2 -term is acting within steady state, that term is zero, due to the Chandrasekhar Theorem (A.16) and because $\Delta Q = 0$ (see (A.4), A=Addendum).

So, $\qquad E(t_s) = E_1(t_s) = E_0 + \int_0^{t_s} N(t)dt. \tag{4.15}$

According to (4.5) and (4.9) during the steady state $t \geq t_s$:

$$E(t) = \int (\alpha T - V^2/(2u^2)) \, d\Omega = 0, \tag{4.16}$$

therefore, during the steady state

$$E(t) = E_1(t_s) = E_0 + \int_0^{t_s} N(t)dt = 0. \tag{4.17}$$

The equation (4.17) means that even if $E_1(t)$ wasn't zero within the transitional period, at the end of that period temperature and velocity have adjusted themselves in such a manner that they have absorbed initial E_0 and made the final value $E_1(t_s) = 0$. That's why when the convection has arrived at the steady state, $t \geq t_s$, the influence of the initial conditions is completely gone.

In other words, *at the steady state, convection is independent from the initial conditions*.

5. Concluding remarks

For the classic convection equations, the density expansion was extended from $\rho = \rho_0(1 - \alpha T')$ to $\rho = \rho_0(1 - \alpha T' + V^2/(2u^2))$.

Using this expansion, the following facts have been proven:

- Steady state convection is independent of any initial conditions;
- During the steady state, global internal convective energy and kinetic energy are equivalent;
- During the steady state, global mass conservation is consistent with extended density expansion and with the conservation of momentum.

These results do not exhaust all the problems of Steady State convection. One important and difficult problem is the estimation of the length of the transitional period between the beginning of convection and the beginning of steady state. It seems to be dependent on initial conditions, on Pr and Ra as well as on the size and form of the container. The equation (4.17) connects a transitional period t_s with the initial condition

$$E_0 = \int (\alpha T_0 - (V_0)^2/(2u^2))\, d\Omega, \tag{5.1}$$

where $T_0(x,z)$ and $V_0(x,z)$ are initial distributions of the temperature and velocity. Note that the equation (4.17) involves only the combination E_0 not the individual temperature and velocity. This means transitional period t_s depends only on the integral combination (5.1) as a whole.

III. NUMERIC EXPERIMENTS ON STEADY STATE CONVECTION

In this chapter results of numeric experiments are presented confirming the fact that steady state convection does not depend on initial conditions.

Calculations are done for two types of initial conditions: vorticity type and temperature type. Thus, the same system of equations is solved numerically several times with different initial conditions to show the same result: steady state convection does not depend on initial conditions.

All calculations presented here are performed for the Standard Convection Conditions, formulated in the Introduction.

At steady state, as it was proven (Schlüter, A., Lortz, D., Busse, F. 1965), the only stable form of convection is straight rolls. In other words, we are dealing with 2D-convection. This important fact was confirmed many times experimentally (Koschmieder E.L., 1993).

Therefore, all our calculations in this chapter are performed for 2D-convection in a 2D-container with all solid boundaries.

In developing methods and algorithms, I will follow the book by Gershuni and Zhukhovitskii (1976).

Below, I will refer to this book as GZ. To help the reader avoid confusion, I have to note some differences in notations with GZ.

Chapter III. **Numeric experiments**

I will use a right coordinate system with x-axis going to the right, z-axis going up, and y-axis going away from the reader beyond the page. GZ use a coordinate system with x-axis going to the right, y-axis going up, and z-axis going towards the reader. In GZ all the functions are independent of z. Here all the functions are independent of y.

For example, the velocity is $V = V(V_x, V_z) = V(V_x(x,z), V_z(x,z))$.

Also, GZ uses β for the thermal expansion coefficient while here, as in all English literature, α is used.

I will solve the convection equations in a 2D rectangular container with the long side L, the short side d, and the aspect ratio $\Gamma = L/d$. $\Gamma = 15$ is chosen to imitate an infinite horizontal layer.

1. The main equations

In this chapter we will study the classical dimensionless system of convection equations with the boundary conditions:

$$\partial V/\partial t + (V\nabla)V/\mathrm{Pr} = -\nabla P + \Delta V + \gamma \mathrm{Ra}T, \qquad (1.1)$$

$$\mathrm{Pr}\cdot\partial T/\partial t + V\nabla T = \Delta T, \qquad (1.2)$$

$$\mathrm{div}V = 0. \qquad (1.3)$$

$$T = 1 \text{ at } z = 0, \quad T = 0 \text{ at } z = 1, \qquad (1.3a)$$

$$V_z = \partial_z V_z = 0 \qquad \text{at } z = 0 \text{ and } z = 1. \qquad (1.4)$$

$$\partial_x T = V_x = \partial_x V_x = 0 \quad \text{at } x = 0 \text{ and } x = \Gamma. \qquad (1.4a)$$

Here $\mathrm{Ra} = g\alpha\Theta d^3/(\nu\chi)$ is the Rayleigh number, $\mathrm{Pr} = \nu/\chi$ is the Prandtl number, and γ is a unit vector along the vertical axis z.

I will call this system the GZ-system (after Gershuni and Zhukhovitskii) to contrast with the LL-system (IV.1.1-1.4) by Landau and Lifshitz.

I will not write the boundary conditions for P because P will be eliminated.

There are several differences of the GZ-system from the analogous LL-systems in Chapter IV.

1) Our constant density ρ_0 was defined at an average constant temperature T_0 and the corresponding hydrostatic pressure satisfying $\nabla P_0 = \rho_0 \mathbf{g}$. In our main equations above, the temperature T is a deviation from T_0 and pressure P is a deviation from P_0. This is different from the previous chapters where T, P were counted from the thermodynamical equilibrium parts (IV.1.5) T_e and P_e:

$$T_e = T_h - z^*(T_h - T_c)/d, \qquad \nabla P_e = \rho_0(1- \alpha T_e)\mathbf{g}.$$

2) Dimensional (with stars) time, pressure, velocity, and temperature expressed by dimensionless (no stars) values are:

$$t^* = t(d^2/\nu), \ \ P^* = P(\rho_0 v \chi/d^2), \ \ V^* = V(\chi/d), \ \ T^* = T\Theta,$$

$$(x^*, y^*, z^*) = (x, y, z)d \tag{1.5}$$

All this is true for the GZ-system.

For the LL-system, which I will use in Chapter IV, we have:

$$P^* = P(\rho_0 v^2/d^2), \ \ V^* = V(v/d), \ \ T^* = T(\Theta \mathrm{Pr}), \tag{1.6}$$

3) The main difference between GZ- and LL- systems in this book is that the former is used as a time-dependent system, while the latter is used as a steady state system.

2. Reduction to new unknowns

We have 4 unknowns (V_x, V_z, P, T) for which we have 4 equations (1.1)-(1.3) and 6 boundary conditions (1.4), (1.4a).

Now we will replace the vector velocity V by the scalar stream and vorticity functions ψ, φ and we'll eliminate the pressure P. That's why we don't need boundary conditions for P.

For any function $f(x,z)$ I use tensor notations $\partial_x f = \partial f/\partial x$, $\partial_z f = \partial f/\partial z$. First, let's introduce a stream function ψ by:

$$V_x = \partial_z \psi, \qquad V_z = -\partial_x \psi. \qquad (2.0)$$

In this particular case indices x and z, applied to V, are indicators of the components of the vector V along the axes of coordinates, not operators of differentiation as it is on the right-hand side of (2.0).

The definition (2.0) *automatically satisfies the equation of mass conservation* (1.3) (equation of continuity). Let i, j, k be unit vectors along the axes x, y, z. To eliminate pressure P from the momentum equation (1.1), take the vector operator **curl** of that equation.

As $\mathbf{curl}V = j\Delta\psi$, then (1.1), (1.2) become

$$\partial_t \Delta\psi + (\partial_z\psi \cdot \partial_x\Delta\psi - \partial_x\psi \cdot \partial_z\Delta\psi)/\mathrm{Pr} = \Delta\Delta\psi - \mathrm{Ra}T_x \qquad (2.1)$$

$$\mathrm{Pr}\cdot\partial_t T + (\partial_z\psi \cdot \partial_x T - \partial_x\psi \cdot \partial_z T) = \Delta T. \qquad (2.2)$$

In the system (2.1) – (2.2) we have replaced three unknown functions, V_x, V_z, P by one function, ψ. The price of this is the rising order of the differential equation (2.1), from 2 to 4 ($\Delta\Delta\psi$). Along with this, the number of boundary conditions must be doubled. Along with this, in the difference equations the number of points approximating the 4-th derivative must be increased.

To avoid these complications, following GZ, I will introduce a new function, the vorticity φ, which besides its ability to simplify our calculations, has a physical meaning as the angular velocity of the elementary liquid volume: $\varphi = -\Delta\psi$.

With this vorticity, the system (2.1) – (2.2) becomes

$$\partial_t\varphi +(\partial_z\psi\cdot\partial_x\varphi - \partial_x\psi\cdot\partial_z\varphi)/\text{Pr} = \Delta\varphi + \text{Ra}T_x, \qquad (2.3)$$

$$\varphi = -\Delta\psi. \qquad (2.4)$$

$$\text{Pr}\cdot\partial_t T +(\partial_z\psi\cdot\partial_x T - \partial_x\psi\cdot\partial_z T) = \Delta T \qquad (2.5)$$

Thus, for the 3 unknown functions φ, ψ, T, we have 3 equations each of 2nd order and 4 boundaries in a 2D-container (upper, lower, left, right). We need $3\times4=12$ boundary conditions.

3. Boundary conditions for the new unknowns

Since the tangent line to the stream function ψ coincides with the velocity vector V, and since $V = 0$ at $z = 0$ or 1, we have:

$$\psi = 0 \quad \text{at } z = 0 \text{ and } 1.$$

Furthermore, since $V_x = 0$ for all x at $z = 0, 1$, and $V_x = \partial_z\psi$, then:

$$\partial_z\psi = 0 \qquad \text{at } z = 0 \text{ and } 1.$$

Thus, for the upper and lower boundaries we have:

$$\psi = \partial_z\psi = 0, \qquad T = 1, \qquad z = 0, \qquad (3.1)$$

$$\psi = \partial_z\psi = 0, \qquad T = 0, \qquad z = 1. \qquad (3.2)$$

Assuming that vertical boundaries are rigid and thermo-insulated, the conditions on them are:

$$\psi = \partial_x\psi = \partial_x T = 0, \qquad x = 0 \text{ and } x = \Gamma. \qquad (3.3)$$

To obtain boundary conditions for the vorticity φ, we need an equation $\varphi = -\Delta\psi$ and some considerations close to the boundaries.

First, start by writing $-\varphi = \partial_{xx}\psi + \partial_{zz}\psi$. Second, consider the expansion of ψ above and close to the lower boundary $z=0$ (h is an arbitrary small distance, which will become a distance between nodes in our discrete system later):

$$\psi(x,0+h) = \psi(x,0) + \partial_z\psi(x,0)h + \partial_{zz}\psi(x,0)h^2/2! + ...$$

This expansion can be rewritten due to (3.1) as:

$$\psi(x,0+h) = \partial_{zz}\psi(x,0)h^2/2! + ... \tag{3.4}$$

Invert this equality to $\partial_{zz}\psi(x,0) = 2\psi(x,0+h)/h^2$. According to (3.2), $\partial_z\psi(x,0) = 0$ for all x. Therefore $\partial_{zz}\psi(x,0) = 0$, from which one can write

$$-\varphi(x,0) = \partial_{xx}\psi + \partial_{zz}\psi = 2\psi(x,0+h)/h^2 + ...\text{or:}$$

$$\varphi(x,0) = -2\psi(x,h)/h^2 . \tag{3.5}$$

A similar condition can be specified for all three other boundaries.

$$\varphi(x,1) = -2\psi(x,1-h)/h^2 . \tag{3.6}$$

$$\varphi(0,z) = -2\psi(h,z)/h^2 . \tag{3.7}$$

$$\varphi(\Gamma,z) = -2\psi(\Gamma-h,z)/h^2 . \tag{3.8}$$

Thus, in order to derive a boundary condition for φ we used the boundary conditions for ψ, $\partial_z\psi$, and for $\partial_{zz}\psi$.

We have a total of how many boundary conditions? We have 12: 4 for ψ : (3.1)-(3.2) + 4 for φ : (3.5)-(3.8) + 4 for T : (3.1)-(3.3).

4. Discretization

Instead of solving a problem within the whole container $\Gamma \times 1$ (Γ is length, 1 is depth) we will seek a solution for the nodes of the net covering the container.

Let N be a number of the vertical nodes, including the border nodes, and $h = 1/(N\text{-}1)$ be the distance (step) between two adjacent vertical nodes. Sometimes calculations are performed with different steps for the vertical and horizontal lines, but here I prefer equal steps for simplicity.

Since $\Gamma = 15$ (see below section 8) will simulate an infinite length for our container, let $M = [\Gamma/h] + 1$ be the number of the horizontal nodes, including the border nodes. Here the brackets mean the nearest whole number. Having defined M and N, we can start a discretization of the variables. To approximate the equations (2.3) – (2.5) let's introduce the space-time net:

$$x_i = h(i-1), \qquad i = 1, 2, \ldots, M, \qquad (4.1)$$

$$z_j = h(j-1), \qquad j = 1, 2, \ldots, N, \qquad (4.2)$$

$$t_n = n\tau, \qquad n = 1, 2, \ldots \; . \qquad (4.3)$$

My notation of the indices differs from that of GZ. They begin their indices with $i = 0, 1, 2\ldots$ but I begin with $i = 1, 2\ldots$ This is because I use the system Mathematica for computer calculations, where any name with index 0 means the so-called "Head" of that name, not a counting index. For any function $f(x,z,t)$ let's denote its value at the nodes of our net as $f_{i,j}^n = f(x_i, z_j, t_n)$.

5. The approximation of equations

Replacing the time derivatives by the one-sided differences and space derivatives by the central differences, the system (2.3) – (2.5) becomes the system of difference equations:

$$\varphi_{i,j}^{n+1} = \varphi_{i,j}^n + \{ \Delta\varphi_{i,j}^n + (Ra/(2h))(T_{i+1,j}^n - T_{i-1,j}^n)-$$

$$(1/(4\text{Pr}\cdot h^2))[(\psi_{i,j+1}^n - \psi_{i,j-1}^n)(\varphi_{i+1,j}^n - \varphi_{i-1,j}^n) -$$

$$(\psi_{i+1,j}^n - \psi_{i-1,j}^n)(\varphi_{i,j+1}^n - \varphi_{i,j-1}^n)]\}\tau , \tag{5.1}$$

$$\Delta\psi_{i,j}^{n+1} = -\varphi_{i,j}^{n+1}, \tag{5.2}$$

$$T_{i,j}^{n+1} = T_{i,j}^n + \left(\frac{1}{\text{Pr}}\right)\{ \Delta T_{i,j}^n - \left(\frac{1}{4h^2}\right)[(\psi_{i,j+1}^{n+1} - \psi_{i,j-1}^{n+1}) \times$$

$$(T_{i+1,j}^n - T_{i-1,j}^n) - (\psi_{i+1,j}^{n+1} - \psi_{i-1,j}^{n+1}) \times (T_{i,j+1}^n - T_{i,j-1}^n)]\}\tau. \tag{5.3}$$

The reason for choosing the system (5.1)-(5.3) among others is due to its extreme simplicity. In the system (5.1) -(5.3), the first and the last expressions are the recurrent formulae, so the functions $\varphi_{i,j}^{n+1}$ and $T_{i,j}^{n+1}$ on the $n+1$ time step are calculated from the functions known from the preceding n-th step and from some functions known from the current $n+1$ step. The only real equation in the system (5.1)-(5.3) is the Poisson equation (5.2). Its solution can be obtained by the method of iterations. The Laplace operator in (5.2) is approximated by the usual scheme:

$$\Delta\psi_{i,j}^n = \frac{1}{h^2} (\psi_{i+1,j}^n + \psi_{i-1,j}^n + \psi_{i,j+1}^n + \psi_{i,j-1}^n - 4\psi_{i,j}^n). \tag{5.4}$$

Equation (5.2) is approximated by the iteration scheme according to the relaxation method:

$$\psi_{i,j}^{n,s+1} = (1 - \omega_0)\psi_{i,j}^{n,s} + \frac{\omega_0}{4} (\psi_{i+1,j}^{n,s} + \psi_{i-1,j}^{n,s+1} + \psi_{i,j+1}^{n,s} + \psi_{i,j-1}^{n,s+1} +$$
$$+h^2 \varphi_{i,j}^{n,s}) \tag{5.5}$$

Here s is the iteration number, and $\omega_0 = \frac{2}{1+Sin(\pi h)}$ is the relaxation parameter (Russel, 1962). The approximating error of this scheme is of the order $O(\tau + h^2)$.

6. The approximation of boundary conditions

The boundary conditions for the stream function and for the temperature are in accordance with section 3:

$$\psi_{i,1} = \psi_{i,N} = 0, \quad i = 1,\dots, M; \qquad \psi_{1,j} = \psi_{M,j} = 0, \ j = 1,\dots, N. \qquad (6.1)$$

$$T_{i,1} = T_{i,2}\,;\ T_{i,M} = T_{i,M-1}, \qquad T_{i,1} = 1,\ T_{iN} = 0, \qquad i = 1,\dots, M. \qquad (6.2)$$

The boundary conditions for the vorticity φ come from (3.5)-(3.8):

$$\varphi_{1,j} = -2\psi_{2,j}/h^2, \qquad \varphi_{M,j} = -2\psi_{M-1,j}/h^2, \qquad j = 1,\dots, N. \qquad (6.3)$$

$$\varphi_{i,1} = -2\psi_{i,2}/h^2, \qquad \varphi_{i,N} = -2\psi_{i,N-1}/h^2. \qquad i = 1,\dots, M. \qquad (6.4)$$

The scheme of the boundary conditions for ψ

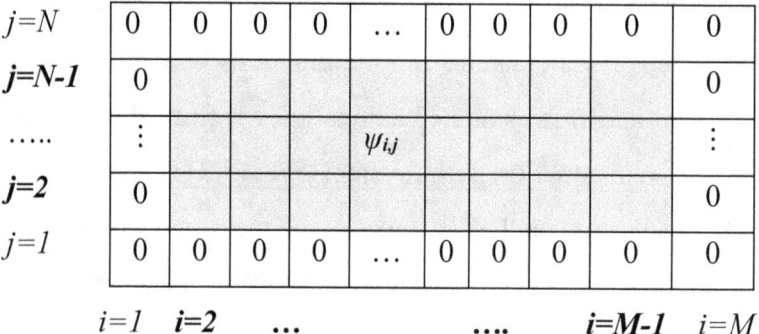

$j=N$	0	0	0	0	...	0	0	0	0	0
$j=N-1$	0									0
.....	$\vdots$				$\psi_{i,j}$					$\vdots$
$j=2$	0									0
$j=1$	0	0	0	0	...	0	0	0	0	0

$i=1 \quad i=2 \quad \dots \qquad \dots. \qquad i=M-1 \quad i=M$

7. Initial conditions

As it was mentioned in the Introduction, this book is about *moderately supercritical* convection or *Koschmieder's convection*:

$$1.2\mathrm{Ra_{cr}} < \mathrm{Ra} < 10\mathrm{Ra_{cr}},$$

where $\mathrm{Ra} = ga\Theta d^3/(v\chi)$.

In order to get such a regime, one must heat a fluid to the given Ra proportional to the temperature Θ. According to the linear theory of convection in the horizontal layer if there are no other disturbing factors then during the slow heating up to Ra_{cr} the fluid will be in the state of mechanical equilibrium with the linear temperature distribution: $T = 1 - z$. Here both T and z are in dimensionless form.

SUBCRITICAL INITIAL CONDITIONS

There are two types of initial conditions that I have used for calculations: *subcritical* initial conditions (this program runs first) and *supercritical* initial conditions (this program will run second). There are many different *supercritical* initial conditions, in whose influence on the steady state we are interested. But there is only one *subcritical* initial condition, $T=1-z$, which we use for every *supercritical* initial condition in the form:

$$T_{i,j} = T_{i,j} = 1 - h(j-1), \quad i=1,\dots, M; \; j=1,\dots, N. \tag{7.1}$$

We also will use *subcritical* initial conditions for ψ and φ :

$$\psi_{i,j} = \varphi_{i,j} = 0, \qquad i=1,\dots, M; \; j=1,\dots, N. \tag{7.2}$$

These conditions play a dual purpose. They reflect mechanical equilibrium before the supercritical regime. They also serve as a cleaning mechanism before each run of the computer program.

The solution of the system (5.1)-(5.3) with subcritical conditions (7.1)-(7.2) is simple. It's (7.1)-(7.2) for any time. On the other hand, the solution of this same system with supercritical initial conditions (specified in the next subsection) is significant because this will provide confirmation that steady state convection does not depend on them.

Chapter III. **Numeric experiments**

SUPERCRITICAL INITIAL CONDITIONS

There are two types of *supercritical* initial conditions I use for calculations: temperature-type and vorticity-type. They are mutually exclusive: if vorticity-type was employed, then the temperature-type wasn't, and vice versa.

As we have seen, *subcritical* initial conditions are placed on each node (i,j) of our net. They remain the same for each run of the program.

In contrast, *supercritical* initial conditions are placed on a small number of nodes and the values are different for each run. Vorticity initial conditions, φ_{ini}, can accept three values:

$$\varphi_{ini} = \varphi_0, \quad \varphi_{ini} = -\varphi_0, \quad \varphi_{ini} = 0 .$$

Similarly, there are three values for the temperature initial conditions:

$$T_{ini} = T_0, \quad T_{ini} = -T_0, \quad T_{ini} = 0 .$$

When $\varphi_{ini} = 0$, then $T_{ini} \neq 0$, and vice versa.

As for ψ, it has just one initial value: $\psi_{ini} = 0$.

The initial vorticity value is placed for every run into the central node with $i = [M/2]$, $j = [N/2]$, where square brackets mean the function Entier, that is the largest integer that is not greater than the value in the brackets.

As for the temperature, initial values are periodically distributed along the line $j = N - 1$ just below the upper boundary. More precisely, the temperature value T_{ini} is filling every 5-th horizontal node.

So, $i = 5, 10, 15,\dots$ are the nodes filled with T_{ini}. All other nodes are filled according to (7.1).

Thus, the (*supercritical*) initial conditions are:

$$\varphi_{i,j} = 0, \text{ if } T_{ini} \neq 0, \quad i=1,\dots, M; \ j=1,\dots, N. \tag{7.3}$$

$$\varphi_{i,j} = \varphi_{ini}, \text{ if } T_{ini} = 0, \quad i= [M/2], \quad j = [N/2], \tag{7.4}$$

$$T_{ini} = 0, \text{ if } \varphi_{ini} \neq 0, \quad i=1,\dots, M; \ j=1,\dots, N. \tag{7.5}$$

$$T_{i,j} = T_{ini}, \text{ if } \varphi_0 = 0, \quad j= N\text{-}1, \quad i = 5, 10, 15,\dots \tag{7.6}$$

8. Preliminary calculations

The calculations presented in this chapter are made for the closed two-dimensional container with dimensionless height 1 and length Γ. That length needs to be long enough for the container to simulate an infinite layer, for which some physical characteristics were found exactly from theory, for example, critical Rayleigh number Ra_{cr} and Neutral curve $Ra(k)$. Experimenters have found that a horizontal container provides the same results as the theory with infinite layer, if the aspect ratio is large enough, $\Gamma \geq 10$.

So, the first thing for calculations was to find optimal Γ, which is large enough, but not so large that it will increase calculation time too much. The sequence of $\Gamma = 1, 5, 10, 15$ was chosen, and for each member of the sequence Ra_{cr} was calculated as a center of fork $Ra_{min} < Ra_{cr} < Ra_{max}$, when one tooth, Ra_{max}, leads to convection development while another tooth, Ra_{min}, makes the fluid stationary. It turns out that all $\Gamma = 1, 5, 10$ make Ra_{cr} significantly larger than theoretical $Ra_{cr}(\Gamma = \infty) = 1708$. $\quad Ra_{cr}(\Gamma = 10) = 1905$. The next try, $Ra_{cr}(\Gamma = 15) = 1708$ has shown that the container with

$$\Gamma = 15 \tag{8.1}$$

is good for imitating an infinite layer.

Chapter III. **Numeric experiments**

In this chapter, I use the term "cell" with a different meaning than in the previous chapters, where the cell was a physical cell in the moving fluid. Below the cell is a mathematical construction within the space net.

As all calculations were done on the space net (4.1), (4.2), the important choice of sizes of the net cell was made taking into account the following:

1) Should the horizontal size of the cell be the same as the vertical h?

2) Should size h remain constant with increasing time?

3) What will be the optimal numeric value for the size h?

For the sake of code simplicity, the answer to the first two questions was Yes. To answer the third question, I needed to find the optimal size h of the cell, so that it is small enough to provide stable calculations approximating the exact solution and large enough to prevent calculations that take too long.

This size h is determined by the number $N - 1$ of the vertical cells in the net:
$$h = 1/(N\text{-}1).\qquad(8.2)$$
There were $N = 13, 17, 21, 51, 76$ with $\Gamma = 15$.

Comparing results (Nusselt number) for different N's, I have found that N =76 provides the optimal precision for $\Gamma = 15$.

Our full space-time net consists of two nets. One is the space net, with which we dealt above, and the other is a time net (4.3).

Now let's find the optimal time step τ for the time net (4.3): $t_n = n\tau$.

The Numerical Analysis Theory says that for stable computations it is necessary that $\tau \leq Ch^2$, where C is some constant, whose value is very difficult to evaluate and it is recommended to find it by trial and error.

In our case it was found by trial and error that

$$\tau = 0.175h^2 \tag{8.3}$$

provides stable calculations for all runs with various initial conditions, both for air and for water. Of course, there exist schemes with variable τ, but for me the simplicity of the algorithm and the program code was the first priority.

9. The algorithm

There are several ways to organize the algorithm for solving the system (5.1)-(5.5). The method I use is **stabilization**.

Its essence is that the calculations of $\psi_{i,j}^n$, $\varphi_{i,j}^n$, $T_{i,j}^n$ from time point t_n to the next t_{n+1} continue until our three functions don't change appreciably. This is considered stabilization.

There are two stabilizations: one that I mentioned above, which is a physical stabilization that is a steady state convection. Another is a numerical stabilization. It requires special care in order to prevent calculations from overflowing. More detail on this will be explained below. The algorithm I used is presented here in a sequence of blocks.

Block A:
1) Choose N, Γ, Pr, Ra.
2) Choose number of iterations, *Niter*, for Poisson equation (5.5).
   Choose δ_ψ - the admissible error in calculating ψ.
   Choose δ_{st} - the admissible error in calculating the beginning of the steady state.
3) Calculate h, ω_0, M.
4) Calculate $\tau = 0.175h^2$.

Block B:
Put the boundary conditions (6.1)-(6.4) into the boundary nodes.

Block C:
Initialize φ, ψ, T. Initialize the iteration number $n = 1$.

Block D:
Increase the iteration number $n = n+1$.
Do one iteration (5.1): $\varphi^s \rightarrow \varphi^{s+1}$

Block E:
Do *Niter* iterations (5.2): $\psi^s \rightarrow \psi^{s+1}$, get $i \times j$ matrix of ψ^{nIter}.

Then do one additional iteration, get $i \times j$ matrix $\psi^{nIter+1}$.

If the difference between the above two is less than δ_ψ, go to Block F, otherwise go to block E.

Block F:
Do one iteration (5.3): $T^s \rightarrow T^{s+1}$

Block G:
Check the stabilization of ψ $(\partial_t \psi \approx 0)$. If ψ is stable do 10,000 iterations, then exit the program, if ψ is not stable go to block **D**. (More details shown in the next section).

10. The criterion for stabilization

As it is clear from the main equations (2.3)-(2.5), our main unknowns are φ, ψ, T. The program will find their values at all nodes (i,j) of the space net at all values of time nodes τn.

Because we are interested in convection at a moderately supercritical regime, $1.2\text{Ra}_{cr} < \text{Ra} < 10\text{Ra}_{cr}$, we know that within that range convection manifests itself as a two-stage process. The first stage is transitional. Depending on initial conditions and on the Prandtl and

Rayleigh numbers, the transitional stage can be of different intensity and of different length of time.

The second stage is stability. Due to the moderately supercritical regime after the transitional stage, stability of convection will come with constant velocity and temperature distribution. Depending on the Prandtl and Rayleigh numbers, these distributions can be different.

For numeric experiments I considered two fluids – air (Pr=0.7) and water (Pr = 7.0), both at Ra = 5,000 and Ra = 20,000.

To estimate the transitional time, I choose ψ out of the three functions φ, ψ, T, because ψ was the most sensitive to changes.

In Figures 10.1 and 10.2 we can see the evolution $\max(\psi_{i,j})$ for two sets of initial conditions:

1) $T_{ini} = 0.1, \varphi = 0$;   Pr = 0.7; Ra = 20,000 (Fig.10.1);
2) $T_{ini} = 0; \varphi = 0.1$; Pr = 7.0; Ra = 5,000 (Fig.10.2).

In both figures see $\max(\psi_{i,j})$ along the vertical axis for dimensionless ψ. Along the horizontal axis see the iteration number n.

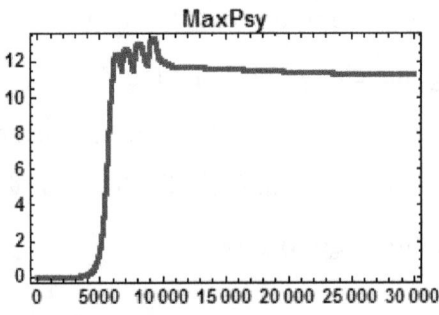

Fig.10.1 $\max\psi(n)$. $T_{ini} = 0.1$; $\varphi_{ini} = 0$; Pr = 0.7; Ra = 20,000.

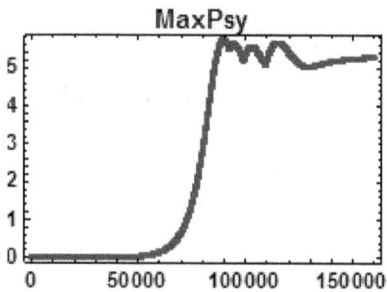

Fig.10.2 max$\psi(n)$. $T_{ini}=0$, $\varphi_{ini}=10.$; Pr $=7.0$; Ra $= 5{,}000$.

Both graphs are qualitatively similar, but quantitatively different.

Both graphs demonstrate the existence of two stages: transitional and steady state.

In the computer program the steady state was determined by the following rules:

- During the initial 2,000 time-steps do nothing.
- Then after every 1,000 time-steps take the last 999 values of the graph of max($\psi_{i,j}$) and approximate these 999 values by the straight line $ax+b$ using the least square method.
- If $a < 10^{-4}$, then $n_{stbl} = n$ - 999 is the time-step number of the beginning of the steady state. If the inequality is not satisfied, continue calculations, go to Block G.

11. The set of initial values

As it was mentioned in section 7, initial values are divided into subcritical and supercritical.

Subcritical values are:

$$\varphi = \psi = 0, \qquad T = 1 - z. \tag{11.1}$$

This is the second solution of the basic equations (2.3)-(2.5), the solution, corresponding to the mechanical equilibrium, when liquid is at rest. The first solution, which we are seeking, has supercritical initial values for φ and T, that I put in the following Table A.

As for ψ, it has initial $\psi = 0$ everywhere.

Thus, (*supercritical*) initial conditions are:

$$\varphi_{i,j} = 0, \text{ if } T_{ini} \neq 0, \qquad i=1,\ldots, M; \; j=1,\ldots, N. \tag{11.2}$$
$$\varphi_{i,j} = \varphi_{ini}, \text{ if } T_{ini} = 0, \quad i= [M/2], \quad j = [N/2],. \tag{11.3}$$

$$T_{ini} = 0, \text{ if } \varphi_{ini} \neq 0, \qquad i=1,\ldots, M; \; j=1,\ldots, N. \tag{11.4}$$
$$T_{i,j} = T_{ini}, \text{ if } \varphi_{ini} = 0, \quad j= N\text{-}1, \qquad i = 5, 10,\ldots \tag{11.5}$$

TABLE A

Init	$\varphi_{i,j}$	$T_{i,j}$
1	10	0
2	-10	0
3	0	0.1
4	0	-0.1

These initials must be supplemented by the parameters Pr -Prandtl number, reflecting liquid viscosity, and by Ra- Rayleigh number, reflecting intensity of heating as it is shown in Table B.

TABLE B

Parameters	Pr	Ra
1	0.7	5,000
2	0.7	20,000
3	7.0	5,000
4	7.0	20,000

So, we have 4 combinations of initials φ and T from Table A for each combination of Pr and Ra from Table B.

Thus, in total we have 4x4=16 combinations, for each of whom I have to run the computer program according to the Algorithm from time=zero until the time when the steady state will be established.

12. Calculation of the results

As it is clear from Tables A and B of section 11, we have 16 results that form four groups shown below in Table C:

For each given combination of Pr and Ra, we have four pairs of initials $\varphi_{i,j}$ and $T_{i,j}$, and for each such pair, $M_\psi(m) = \max[\psi_{i,j}]$ is calculated when n_{stbl} has been reached for the steady state ($m=1,2,3,4$).

Those four $M_\psi(m)$'s are compared by the standard statistical analysis in order to find the relative standard deviation (RSD), which I express percentagewise:

$\varepsilon_\psi = [\text{SUM}_{m=1\ to\ 4}(M_\psi(m) - M_{avr})^2/4]^{1/2}/M_{avr})100\%$,

$M_{avr} = [\ M_\psi(m=1) + M_\psi(m=2) + M_\psi(m=3) + M_\psi(m=4)\ ]/4$.

ε_φ and ε_T are calculated in the same manner, beginning with $M_\varphi(m)=\max[\varphi_{i,j}]$ and $M_T(m)=\max[T_{i,j}]$.

$In=(In1,\ In2,\ In3,\ In4),\ \psi = 0.$

In1= $\{\varphi_{i,j} = 10,\ i=[M/2], j=[N/2],\ T_{ini}=0,\ i=1,...,M;\ j=1,...,N.\}$

In2= $\{\varphi_{i,j} = -10,\ i=[M/2], j=[N/2],\ T_{ini}=0,\ i=1,...,M;\ j=1,...,N.\}$

In3= $\{\ \varphi_{i,j} = 0,\ i=1,...,M;\ j=1,...,N.\ T_{ini}=0.1,\ j=[N/2],\ i=5.10,...\}$

In4= $\{\ \varphi_{i,j} = 0,\ i=1,...,M;\ j=1,...,N.\ T_{ini}=-0.1, j=[N/2],\ i=5,10,...\}$

TABLE C

	Parameters		4 initials	results of 4 initials		
m	Pr	Ra	In	ε_ψ	ε_φ	ε_T
1	0.7	20,000	$In1$	2.6%	1.8%	0.24%
2	7.0	20,000	$In2$	0.6%	0.12%	0.11%
3	0.7	5,000	$In3$	1.4%	1.08%	0.39%
4	7.0	5,000	$In4$	0.4%	0.004%	0.01%

Thus, our calculations show that at the steady state each of the three $Max(\psi)$, $Max(\varphi)$, $Max(T)$ does not depend on various types of initial conditions within statistical relative error less than 3%.

This confirms the theoretical conclusion in Chapter II regarding the independence of steady state convection from initial conditions.

The last conclusion is not the first of its kind.

In the paper by T.D. Foster, (1969, Fig. 4) there are two graphs showing the development of the Nusselt number with time, starting with two different initial conditions and merging into one resulting line. Below I have copied Foster's graphs.

In both cases with Ra = 16Ra$_{cr}$ and with Ra = 4Ra$_{cr}$ one can clearly see that the results are independent of initial conditions.

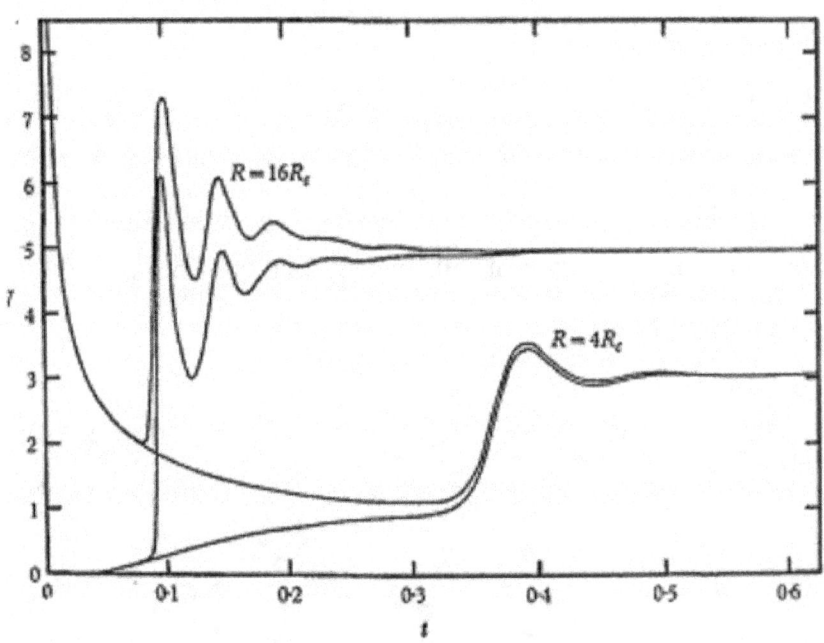

RE 4. Time development of Nusselt number starting from a sudden change temperature and an infinitesimal disturbance.

13. **Illustrations of calculations**

This section consists of illustrations for numerical results and of rolls' internal structure.

Below we see the general picture (after Koschmieder, 1993) of the convection cells of diameter λ and of thickness d formed by adjacent rolls. The small arrows indicate the direction of the rotations. The cell is determined by two adjacent rolls, where fluid **rises** between them.

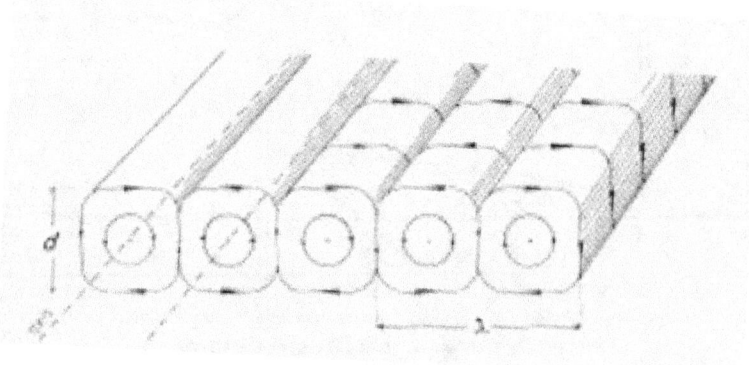

Fig.13.1. Convection rolls.

So, from right to left 1st and 2nd rolls compose the cell as well as 3rd and 4th. 5th is just a roll.

Except for a schematic in Fig.13.1, most illustrations in this section are presented in the form of the Contour Plots.

Contour Plots (also called Level Plots) are a way to show a three-dimensional surface $\varphi = \varphi(x,z)$ on a two-dimensional plane (x,z).

A contour plot is composed of several Contour lines. Each line is a 2D-graph of the function $\varphi(x,z) = C$ or $y = f(x,C)$, where f is inverse of φ, C is a constant.

In other words, a Contour curve is an intersection curve for the horizontal plane $z = C$, cutting the surface $\varphi(x,z)$. I am using the standard program for plotting Contour plots, where number of C's are chosen manually or automatically depending on the height of the surface and in order to make the plot more aesthetical.

The schematics on Fig.13.1 corresponds to the contour plot Fig.13.2 from my calculations showing vorticity $\varphi=\varphi(x,z)$ in container of height 1 and of length 15: The light areas are the rolls rotating clockwise.

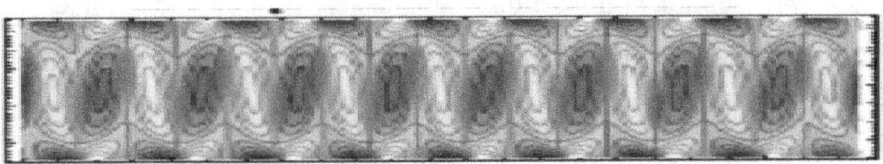

Fig. 13.2. The contour plot of vorticity. $\varphi=\varphi(x,z)$. Steady state.

The dark areas are the rolls rotating counter-clockwise. The cell is composed by the light roll on the right side and the dark roll on the left side. (See details below.)

The vorticity is a vector, its magnitude is φ. The direction of the vorticity is given by the "right-hand" rule. So, the **counter-clock** rotation of the convection roll means that the direction of the vorticity is out of this page **towards the reader**. This is the negative direction of the axis y of our coordinate system as it was defined in the introduction to this Chapter III: "I will use a right coordinate system with x-axis going to the right, z-axis going up, and y-axis going away from the reader beyond the page".

The counter plot in Fig.13.2 was calculated with the computer program Mathematica-8, according to which *larger values are shown lighter.* This means that the vorticity with lighter color on Fig.13.2 represents a vector directed from the reader, while dark color represents a vector directed to the reader. Closed contour lines are the lines of constant vorticity, the lines along which the liquid rotates. Closed contour lines on the light area represent clockwise rotation.

As it was mentioned in section 4 of Chapter III, in 2D box convection is determined by three functions: vorticity $\varphi(x,z)$, temperature $T(x,z)$, and stream function $\psi(x,z)$. For numeric calculations these functions were defined on the space-time net:

$$x_i = h(i-1), \qquad i = 1, 2,\ldots, 1126, \qquad\qquad (13.1)$$

$$z_j = h(j-1), \qquad j = 1, 2,\ldots, 76, \qquad\qquad (13.2)$$

$$t_n = n\tau, \qquad n = 1, 2,\ldots \;, \qquad\qquad (13.3)$$

where $h=1/75 = 0.01333333$, $\tau = 0.175h^2 = 0.00003111111$.

Functions $\varphi_{i,j}^n = \varphi_n(x_i, z_j)$, $T_{i,j}^n = T_n(x_i, z_j)$, $\psi_{i,j}^n = \psi_n(x_i, z_j)$ are calculated on the space net $\{x_i, z_j\}$ at each time-step t_n with $n = 1$, 2,3,...100,000 , ...

All calculations are beginning with initial values. All initial values are broken into two major groups: Vorticity group and Temperature group. The initial stream function $\psi(x,z)$ is always zero for both groups.

Below, the contour plots will be supplemented by regular 2D graphs for two curves: $M(\varphi_n) = \max\limits_{i,j} \varphi_n(x_i, z_j)$, $M(T_n) = \max\limits_{i,j} T_n(x_i, z_j)$.

In the next page I have shown contour plots first for initial $\varphi(t=0)$, $T(t=0)$ and then for $t=t_n$, $\varphi_n(x,z)$, $T_n(x,z)$ and corresponding graphs of $M(\varphi_n)$, $M(T_n)$ for three time-steps: one is close to initials

$$\varphi(i = M/2, j=N/2) = -0.1,$$

second is intermediate during transition, third is during the steady state. All of them are calculated for air ($Pr = 0.7$) at $Ra = 20,000$.

While Fig.13.2 was in scale (more or less, almost), all the following contour plots will be compressed horizontally and stretched vertically.

Vorticity initials

This is schematics for air ($Pr = 0.7$) at $Ra = 20,000$.

Positive value means clock-wise rotation according to the "right-hand rule" and axis y directed from the reader beyond the page.

$$\varphi(i = M/2, j=N/2, t=0) = \textbf{10.0}$$

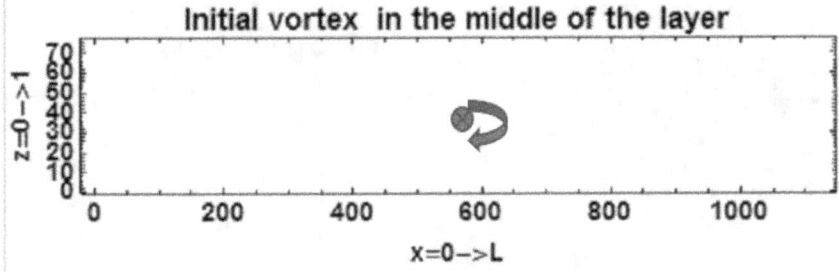

Fig. 13.3. Initial vorticity $\varphi=\varphi(x,z)$.

If φ would be negative, the rotation would be in the opposite direction.

The shown vortex is pointwise nevertheless it is rotating with a rate of 0.13 revolutions per minute = 47 degree per minute. Very slow.

Two pictures below show vortex and temperature distributions after 100 time-steps. We can see very primitive vortex and purely linear temperature distribution from 1 on the bottom to 0 at the top without any trace of convection. The vertical and horizontal scale showing number of space units $h = 0.0133333$.

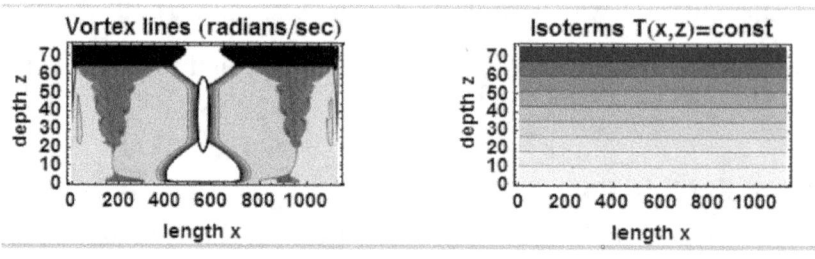

Fig. 13.4. Vorticity and temperature after 100 time-steps.

Next we see pictures after 70,000 time-steps. Convection begins at the center of container and will propagate to full volume.

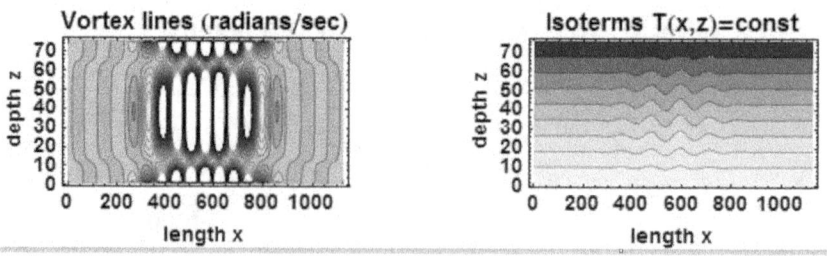

Fig. 13.5. Vorticity and temperature after 70,000 time-steps.

Next we will see pictures at the steady state.

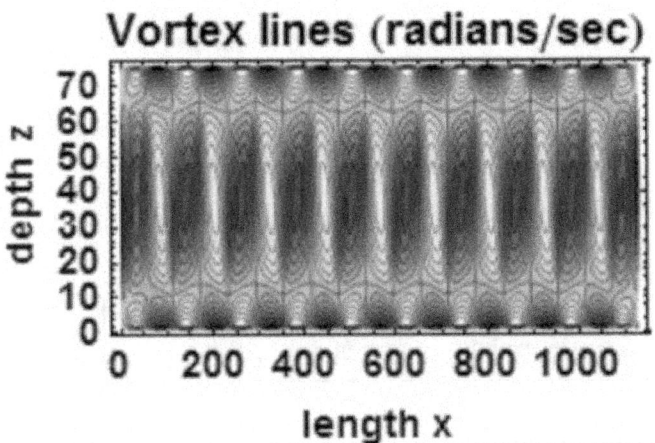

Fig. 13.6. Vorticity after 180,000 time-steps.
9 cells filling container. Steady state.

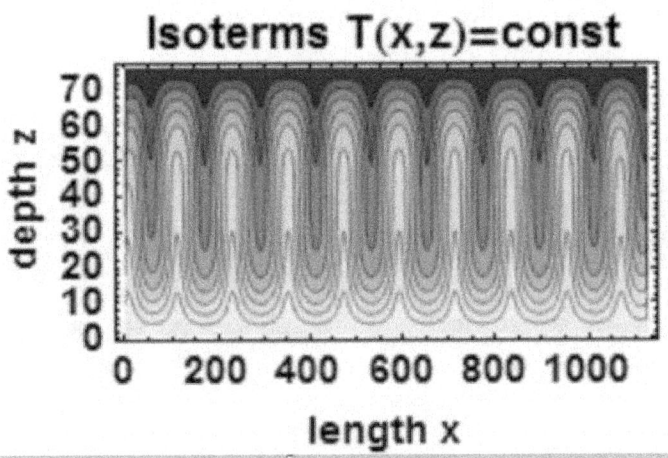

Fig. 13.7. Temperature after 180,000 time-steps.
9 cells filling container. Steady state.

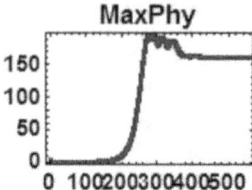 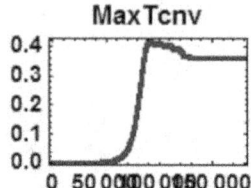

Fig. 13.8. Vorticity and temperature in 3 periods:
beginning, transition and steady state.

Next we have illustrations for initial vortex in opposite direction to the previous case, namely $\varphi(i = M/2, \, j = N/2, \, t = 0) = -10.0$

Nothing will be different with Fig.13.4 except I put the contour plot for the stream function along with the vorticity contour plot.

Result: after first 100 steps of the algorithm the streamlines are well developed at the center of the container while vorticity is in a primitive stage.

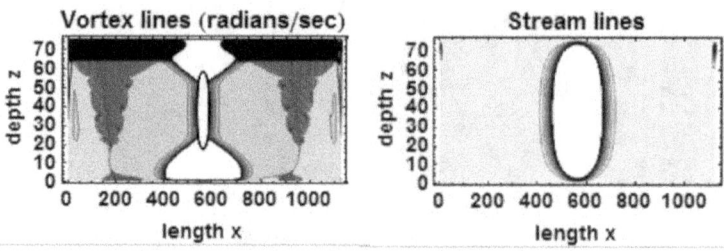

Fig. 13.9. Vorticity and streamlines after 100
time-steps. Steady state.

The following pictures reflect steady state vorticity and temperature.

Here are the results for $\varphi(t{=}0) =$ **-10.0** as compared to $\varphi =$ **+10.0** .

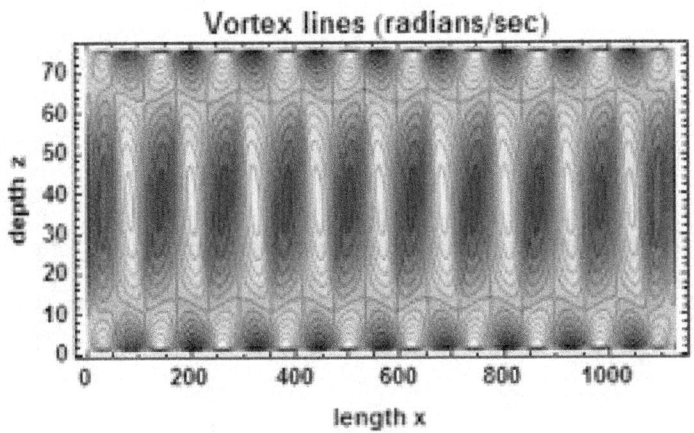

Fig. 13.10. Vorticity after 180,000 time-steps.
Steady state.

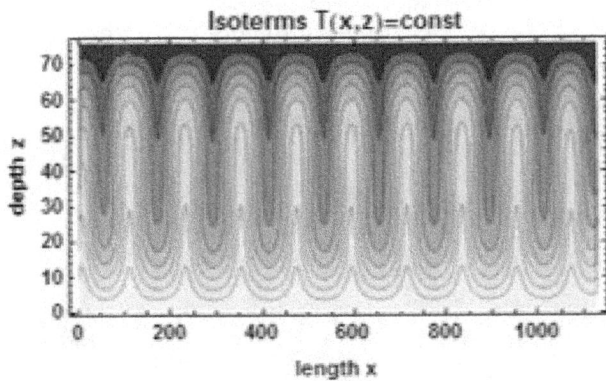

Fig. 13.11. Temperature after 180,000 time-steps.
Steady state.

NO VISIBLE DIFFERENCES FROM Fig.13.6, 13.7.

Temperature initials

Now I will repeat for the temperature initials $T(t=0)$ everything what was shown for vorticity initials.

Positive case

In the next pages I will show contour plots first for initial $T(t=0)$ and then for $t=t_n$, and corresponding graphs of $M(\varphi_n)$, $M(T_n)$ for three time-steps: one is close to initials ($\varphi_0 = 0$,)

$$T_{i,j} = T_{ini} = 0.1, \quad j = N\text{-}1, \quad i = 5, 10, 15,\dots ,$$

the second is intermediate during transition, the third is during the steady state. All of them are calculated for air (Pr = 0.7) at Ra = 20,000.

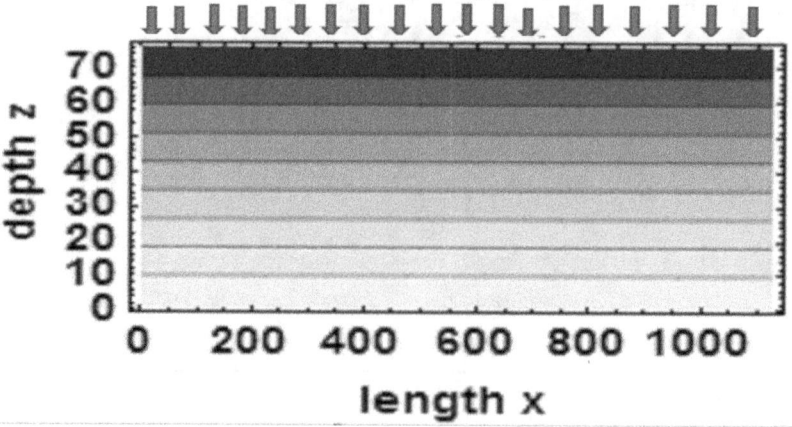

Fig. 13.12. Initial temperature. Positive case.

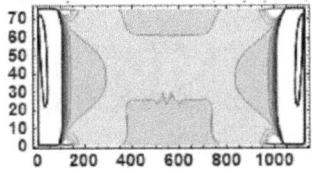

Fig. 8a. Photograph of the structure just at the start of the instability

Fig. 13.13. Numerical (left after 100 initial time steps) and experimental (right, after Berge) images of the beginning of convection. Both show the formation of cells at the vertical boundaries.

These pictures confirm that in the absence of initial vorticity the convection begins at the side walls. As it was noticed by R.P. Behringer (1985) "…the vertical walls exert extra drag on the fluid which must be overcome by the buoyancy force for convection to begin."

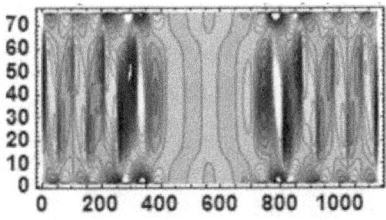

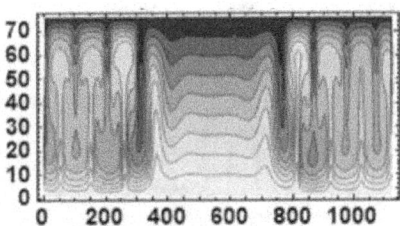

Fig. 13.14 Vorticity lines and temperature isoterms after 30,000 steps of the algorithm. Vertical and horizontal number are space steps h.

These pictures show that convection begins at the side walls and propagate to the center.

Below are the pictures taken at steady state.

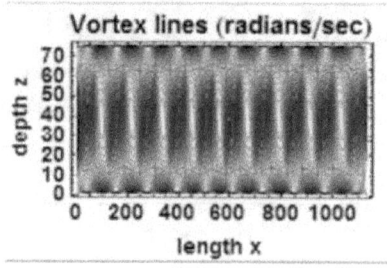 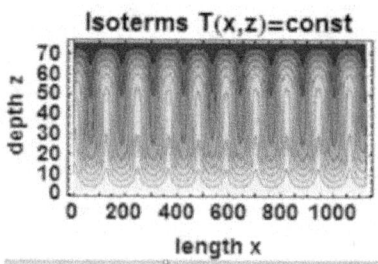

Fig. 13.15 Vorticity lines and temperature isotherms
after 190,000 steps of the algorithm.
9 cells fill the container. Steady state.

These pictures show the cells are formed throughout the whole

container and will remain the same.

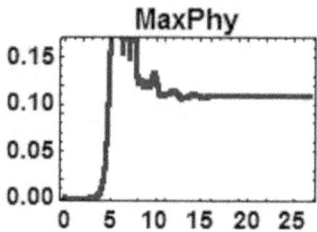 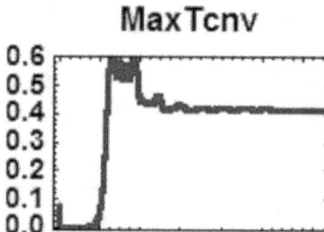

Fig. 13.16. These pictures show three stages of
convection: the initial period, transition, and
steady state.

Negative case

Next we have illustrations for initial temperature with opposite sign to the previous case, namely

$$T_{i,j} = T_{ini} = \mathbf{-0.1}, \quad j = N\text{-}1, \quad i = 5, 10, 15, \dots,$$

This is schematics for air (Pr = 0.7) at Ra = 20,000.

Initial temperature $T(x_n, z{=}1) = -0.1;$ *Schematic, not in scale.*

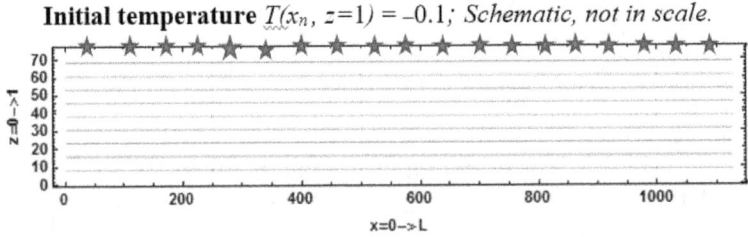

Fig. 13.17. Initial temperature. Negative case.

Below we compare vorticity distribution created with two different initial temperatures. On the left is the case with initial positive temperature. On the right is the case with initial negative temperature. Even at this early stage vorticity distribution in both cases is the same.

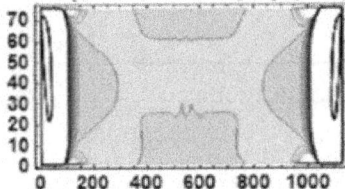 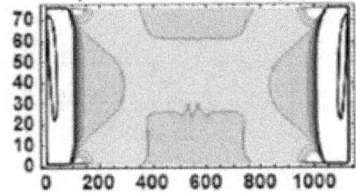

Fig. 13.18. Comparison of vorticity development after 100 initial time steps for two initial temperatures.

However, much later at the steady state we can observe that the distributions of vorticity and of the temperature being the same have shifted relative to one another.

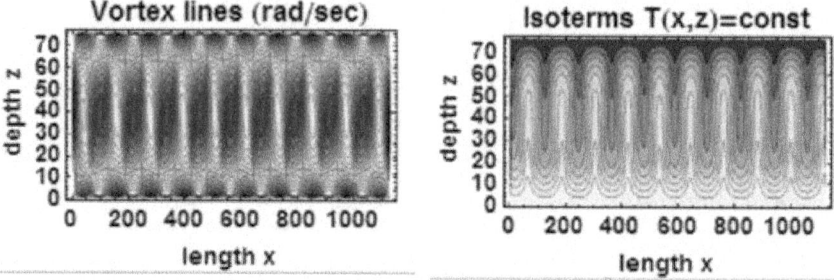

Fig. 13.19. Vorticity and temperature after 140,000 time-steps for positive initial temperatures.

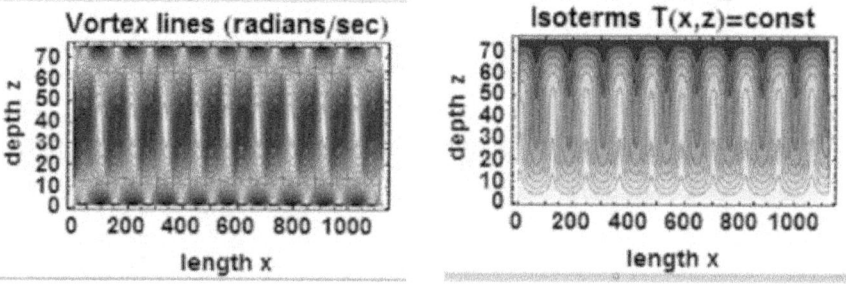

Fig. 13.20. Vorticity and temperature after 240,000 time-steps for positive initial temperatures.

Rolls' internal structure

During the steady state convection cells fill up completely the container with rigid boundaries.

BOUNDARY LAYER

However, 'completely' should be understood conditionally. There are **boundary layers** formed along upper and lower boundaries as we can see in the following Fig. 13.21.

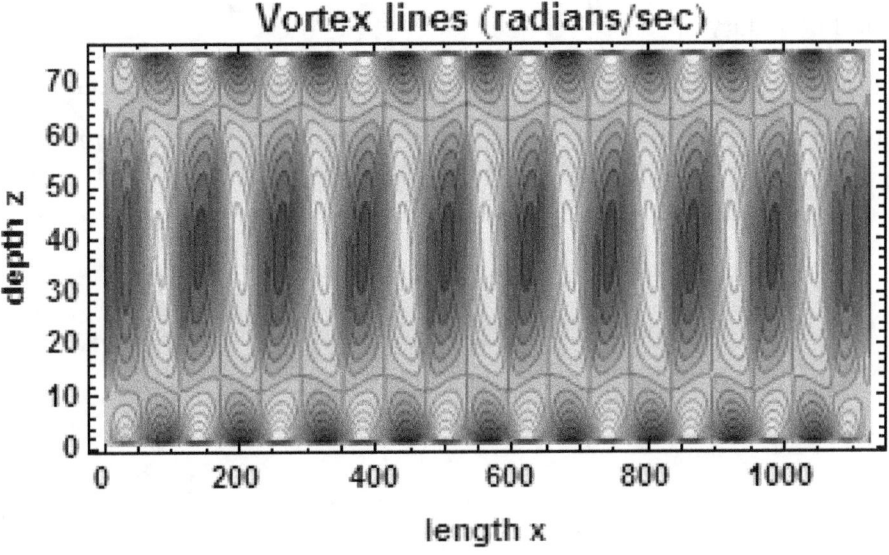

Fig. 13.21. Vorticities and their boundary layers.
Steady state.

There is a continuous curve separating each layer from the main volume filled by vorticity cells. Boundary layers have cellular structure similar to the corresponding vorticity cells. The boundary cell corresponding to the volume cell has antisymmetric construction: if the volume roll has clock-wise rotation (light color) then the corresponding boundary roll has counter-clock rotation (dark color).

However, boundary rolls don't rotate at all, because they belong to the boundary layer. Nevertheless, the fluid within the boundary rolls is moving half a 'roll' following to the right-hand rule but stopping at the solid boundary.

Boundary layer starts to form at the very beginning of convection.

It is present on all the above illustrations from Fig.13.2 through Fig.13.25.

IDENTITY OF CELLS

As for the cells, we can see that they are the same except small differences in the leftmost and rightmost cells. All cells have straight vertical boundaries.

INCLINATION OF ISOLINES

However, the plain, which divides each 'light' roll into symmetrical halves is inclined from lower right to upper left side for clock-wise rotation and for the 'dark' rolls dividing the plane is inclined from lower left to upper right side. The inclination is better visible on the enlarged portion of Fig. 13.22, when it is stretched closer to scale:

Fig. 13.22. Vorticity shows inclination of isolines in opposite directions for light and dark rolls.

TEMPERATURE

Below in Fig. 13.23 the cells are shown as 'light' and 'dark' pairs of rolls. The individual cell has a dark roll on the left, and adjacent to it is a light roll on the right. This is because light rolls rotate counter-clock wise while dark rolls rotate clock-wise, so the fluid rises in between, which is the main ingredient of the definition of a cell.

In both light and dark rolls, isotherm lines are going monotonously from the higher value on the bottom to the lower value at the top.

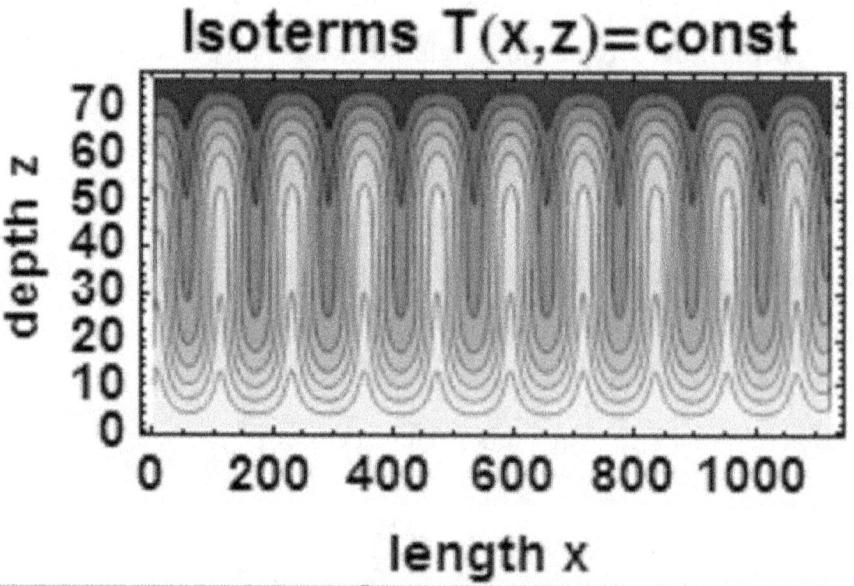

Isoterms T(x,z)=const

Fig. 13.23. Temperature distribution is periodic horizontally and monotony vertically. Steady state.

STREAM LINES

For any function $f(x,z)$ I use tensor notations $\partial_x f = \partial f / \partial x$, $\partial_z f = \partial f / \partial z$. Indices x and z, applied to V, are indicators of the components of the vector V, not operators of differentiation ∂x or ∂z.

Stream lines are level lines of the stream function $\psi(x,z) = constant$. The stream function is defined by the velocity functions:

$$V_x = \partial_z \psi, \qquad V_z = -\partial_x \psi.$$

This definition automatically satisfies the equation of mass conservation (equation of continuity).

The velocity vector is a tangent vector to the stream line.

Below, in Fig. 13.24 the cells are shown as 'light' and 'dark' pairs.

The lighter area shows the larger value. Light areas are the rolls rotating

clock-wise. So, a cell is composed of two adjacent rolls (lighter is right

roll).

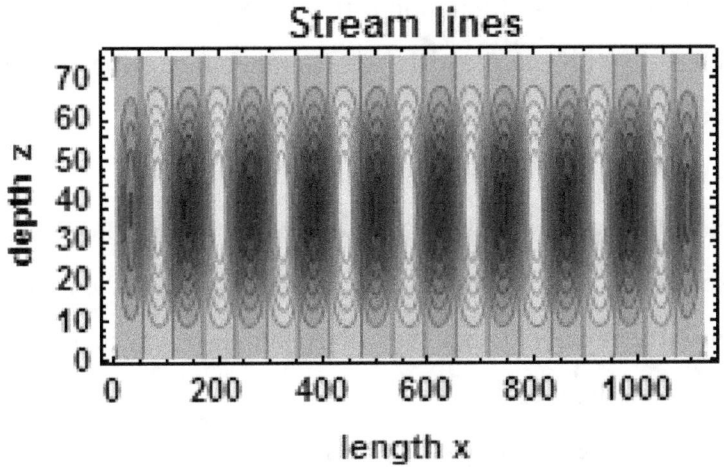

Fig. 13.24. Stream lines show that velocity distribution is periodical horizontally and symmetrical vertically. Steady state.

Plot on Fig.13.24 is not in scale. Its part closer to scale is Fig.13.25.

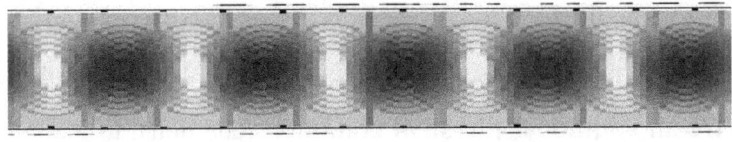

Fig. 13.25. This is a part of the previous figure.

IV. WAVE NUMBER SELECTION AT STEADY STATE CONVECTION

Two fundamental contradictions between experimental convection and convection theories have been described by Koschmieder (1993) in his famous book "Benard Cells and Taylor Vortices".

The first fundamental contradiction is that the experiments show an increasing cell size with increased heating temperature, whereas the theories predict the opposite.

Earlier (1988), I. Catton came to the same conclusion:

"The characteristic wave number is a decreasing function (larger wavelength) of the Rayleigh number in contrast with many theoretical efforts, which have predicted that the wave number is an increasing function of Rayleigh number."

The second fundamental contradiction is that while the experiments show a unique dependence of cell size on the heating temperature, the theories show a whole area filled with various cell sizes for one temperature. That means non-uniqueness.

The main idea of this chapter is to show that initial and boundary conditions are not enough for determining a unique solution of convection equations. This is because the steady state convection possesses internal structure - the cells. To make a unique solution we must have an additional **physical** condition (additional to the initial and boundary conditions) determining the size of the cells. This additional condition is the Principle of Maximum Entropy Production.

As I. Catton (1988) remarks: "…it appears that some governing principle(s) must exist, which forces uniqueness or closure."

Chapter IV. **Wave number selection**

1. The main equations

In this chapter, a simple model based on the principle of maximum entropy production is suggested, which explains both of the experimental features in cell size behavior.

I begin by rewriting the classical dimensionless convection equations for an incompressible fluid ($\partial_t V = \partial V / \partial t$, $\partial_t T = \partial T / \partial t$) :

$$\partial_t V + (V \nabla) V = -\nabla P + \Delta V + \mathrm{Ra} T \gamma, \qquad (1.1)$$

$$\mathrm{Pr}(\partial_t T + V \nabla T) = \Delta T + V_z, \qquad (1.2)$$

$$\mathrm{div} V = 0, \qquad (1.3)$$

where V is the fluid velocity, T is the temperature, P is the pressure, and $\gamma(0,0,1)$ is a unit vector along the vertical axis z. The actual temperature distribution between the plates consists of the sum of two functions. One of them is the conductive temperature,

$$T_{cnd} = 1 - z, \qquad (1.4)$$

and the other is the convective temperature, T. Temperature T is the deviation from T_{cnd} caused by convection. Similarly, V is the deviation from the mechanical equilibrium $V = 0$, while P is the deviation from the hydrostatic pressure P_e due to gravitation and the equilibrium temperature.

All three V, T, P are written in dimensionless form in (1.1) - (1.4).

In dimensional units we have:

$$T_e = T_h - z^*(T_h - T_c)/d, \qquad P_e = -\rho_0(1 - \alpha T_e)g. \qquad (1.5)$$

Dimensionless time, pressure, velocity, temperature, and length are

$$t = t^*/(d^2/v), \; P = P^*/(\rho_0 v^2/d^2), \; V = V^*/(v/d), \; T = T^*/(\Theta \mathrm{Pr}),$$

$$(x,y,z) = (x^*,y^*,z^*)/d, \qquad (1.6)$$

and letters with a star correspond to dimensional values; d is the thickness of the fluid layer.

2. Boundary conditions

The system (1.1) - (1.3) is a classical model describing convection in a horizontally infinite fluid layer between two parallel plates (solid boundaries), at which constant temperature, pressure, and no-slip velocity is maintained:

$$T = P = 0, \qquad\qquad z = 0 \text{ or } 1, \qquad\qquad (2.1)$$

$$V_z = \partial V_z/\partial z = 0, \qquad z = 0 \text{ or } 1. \qquad\qquad (2.2)$$

The system with free boundaries will have the same equations, the same boundary conditions (2.1), but different (2.2):

$$V_z = \partial^2 V_z/\partial z^2 = 0, \qquad z = 0 \text{ or } 1. \qquad\qquad (2.3)$$

For the cell $\Omega = \{ -\pi/k \leq x \leq \pi/k, \; 0 \leq z \leq 1 \}$, the conditions on the left and right boundaries are:

$$V_x(x,z) = \partial T(x,z)/\partial x = 0, \qquad x = -\pi/k \text{ or } \pi/k. \qquad (2.1.F)$$

3. Entropy production

Let's begin by writing the internal energy balance in terms of entropy (Landau & Lifshitz, 1987, §49):

$$T(\partial_t s + V\nabla s) = \Delta T, \qquad (3.1)$$

Here I write the temperature without splitting it into conducting and convective parts.

To get an expression for the entropy production, I rewrite (3.1) as

$$\partial_t s = -V\nabla s + \operatorname{div}\nabla T/T, \qquad (3.2)$$

Now I integrate (3.2) over the liquid layer Ω (see Addendum).

The result is an Entropy Production:

$$W = d/dt(\int s\,d\Omega) = \int (\nabla T/T)^2 d\Omega , \qquad (3.3)$$

Let's simplify the right part in (3.3), using it to the most characteristic conditions at which most experiments are performed. Usually it is around room temperature 293 ^{0}K, with the lower boundary hotter than the upper one by 1 or 2 degrees. I write the dimensional temperature at the bottom as $T^* = 295$.

Let $d = 1$cm be the typical height of the fluid layer.

$$T^* = 295 - 2z^* + T'(x^*, y^*, z^*), \qquad \nabla T^* = -2 + \nabla T'. \qquad (3.4)$$

Here T' is a small dimensional disturbance of the temperature, which we study along with V. Then (3.3) is

$$W = \int (-2 + \nabla T')^2/(295 - 2z^* + T')^2 d\Omega. \qquad (3.5)$$

Later on, I will try to find the wave number k, which provides the maximum of W. That k will determine the size $\lambda = 2\pi/k$ of the convective cell. When we seek, for example, the maximum of the function

$F(k)=a+bf(k)$ with constants a and $b>0$, then we know that $maxF(k)=maxf(k)$. In view of this, any constants added to W or any positive constant multiplied by W are unnecessary in finding the k for maximizing $W(k)$. Therefore, such terms will be discarded.

In the denominator of (3.5), the temperature 295 is much larger than the rest, so the rest of the denominator terms will be discarded.

In the numerator, the constant term $(-2)^2$ is also discarded, as well as the integral of $\nabla T'$ since the convective temperature T' is zero on both boundaries. Thus, after omitting all additive and multiplicative constants that don't contribute to k, we are left with:

$$W = \int(\nabla T)^2 d\Omega, \quad \Omega = \{ -\pi/k \leq x \leq \pi/k,\ 0 \leq z \leq 1\}. \tag{3.6}$$

From now on, all three W, T, Ω are considered dimensionless.

Let us call W *the convective entropy production* or simply *entropy production.* It will be used for many fluids. A few exceptions are liquid helium and alike, whose temperature is close to absolute zero.

4. The W-model

In this chapter I will consider 2D-convection with simple rolls.

A cell is composed of two adjacent rolls with fluid rising in the central area and sinking at the sides. Moreover, all the parameters and functions will be constant along the direction parallel to the roll's axes. In other words, convection here is two-dimensional. The reason for the 2D-case is the work of Schlüter, Lortz, and Busse (1965), where it was shown that only 2D-rolls are stable in the steady state, not 3D-cells.

The question is: where are the cell boundaries situated? In other words, what is the size of the cell, λ, or its wave number $k=2\pi/\lambda$?

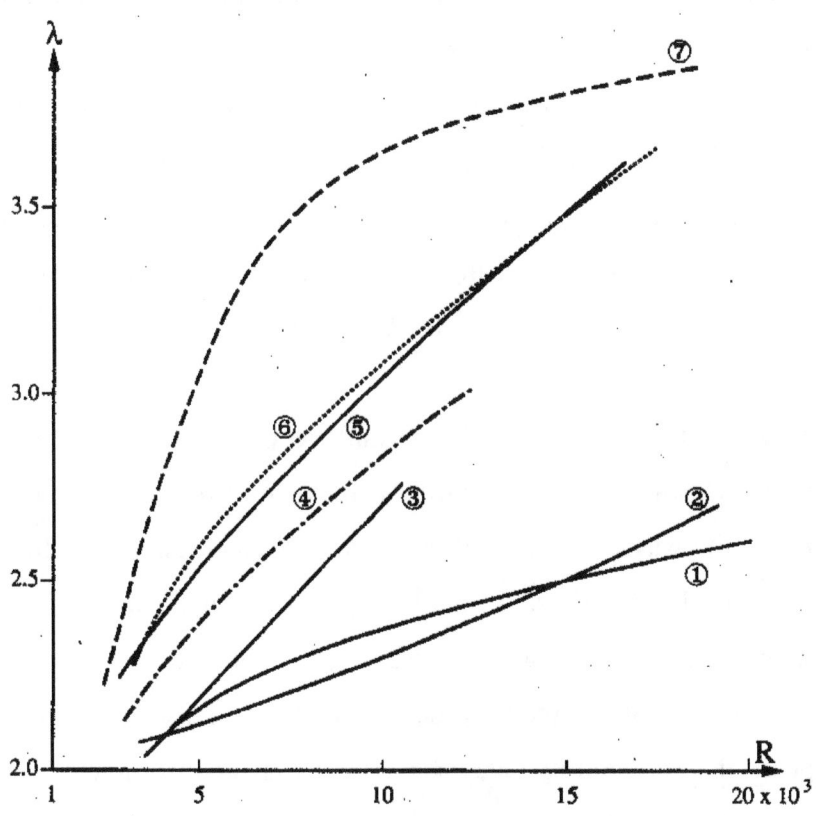

Fig.4.1 **Koschmieder's curves from experiments:**
(1) Silveston (1958), silicone oil, Pr ☼ 3000.
(2) Krishnamurti (1970a), silicone oil, Pr ☼ 870.
(3) Koschmieder and Pallas (1974), silicone oil, Pr☼950.
(4) Martinet et al. (1984), air, Pr ☼ 0.6.
(5) Willis et al. (1972), water, Pr ☼ 6.
(6) Farhadieh and Tankin (1974), water, Pr ☼ 6.
(7) Willis el al. (1972), air, Pr ☼ 0.6.
See references and details in Koschmieder's book (1993).

Chapter IV. **Wave number selection**

If there is no special conditions for k, then any k is acceptable while it is within the Neutral Curve of linear theory or if it is within some balloon (k, Ra), such as the 'Busse Balloon' (Busse, 1967) in the non - linear case. The usual answer to the posed question is that k should be determined from the stability conditions. There is a problem with this. Results of stability analysis almost always contradict the experiments (Fig.4.1), which show decreasing k (increasing λ) with increasing Ra, while the stability analysis shows the opposite.

In Buell and Catton (1986), the authors cite experiments in air (Pr=0.7) performed by Brown, Gille, and Pocheau *at al*, all of whom found that the wave number will decrease with (increasing) Ra. In his paper (Catton,1988) writes: "In examining experimental data it is found that a large amount of data falls outside of the stability envelope and virtually no data falls in the right half <of the balloon Busse – J.K.>. Thus, it appears, that the stability analysis of Busse may be incomplete..."

Fig.4.1 shows the results copied from Koschmieder's (1993) book, showing 7 experiments made by independent researchers with different apparatus and with different fluids. All 7 results show a similar behavior of the cell size: $d\lambda/dRa > 0$, which means **the cell size increases when Ra is increased**. Note, all 7 experiments are done within *moderately supercritical* regime defined in the Introduction.

We need an **independent physical condition** that determines the diameter λ of a cell or a condition for the wave number $k=2\pi/\lambda$. Such a condition is given by the *Principle of Maximum Entropy Production*:

$$max_k\,[W(k)] = max_k\,[\textstyle\int (\nabla T)^2 \; d\Omega], \qquad (4.0)$$

Let us call the *W-model* (for steady state convection) the system of equations together with the Principle of Maximum Entropy Production and together with the usual boundary conditions:

$$(V\nabla)V = -\nabla P + \Delta V + \text{Ra}T\gamma, \tag{4.1}$$

$$\text{Pr}(V\nabla)T = \Delta T + V_z, \tag{4.2}$$

$$\text{div}V = 0, \tag{4.3}$$

$$max_k [W(k)] = max_k [\int(\nabla T)^2 \, d\Omega], \tag{4.4}$$

$$V = T = P = 0, \quad z = 0 \text{ or } 1, \tag{4.5}$$

$$\partial V_z/\partial z = 0, \quad z = 0 \text{ or } 1 \tag{4.6}$$

or at the free boundaries:

$$V_z = \partial^2 V_z/\partial z^2 = 0, \quad z = 0 \text{ or } 1. \tag{4.7}$$

$$V_x(x,z) = \partial T/\partial x = 0, \quad x = -\pi/k \text{ or } \pi/k. \tag{4.8}$$

As we are considering a two-dimensional case, all functions depend on space variables, x, z, and on the parameters k, Pr, Ra .

5. **Earlier attempts to use the W-model**

The Principle of Maximum Entropy Production has been known since long ago (Swenson, 1988), but in convection theory it is rarely used. First, as a stability criterion it was formulated exactly as (4.0) in the paper of Malkus and Veronis (1958). However, in their stability analysis the wavenumber increases with increasing Rayleigh number. This was the wrong conclusion, contradicting the experiments. What is important is that they formulated the concept of convective entropy exactly as in (3.6), albeit not acknowledging its thermodynamical contents: "We prescribe no (thermodynamic) role for this 'entropy'." They connected their entropy to the heat flux and called that connection 'the equation for the balance of entropy, associated with the disturbance T'. We can see such a balance from the Second Theorem of Balance (II.2.2), from which it follows:

$$\int V_z T d\Omega = \int (\nabla T)^2 d\Omega. \qquad (5.1)$$

This is a rather general equation. On the left side we have a heat flux through the fluid layer; on the right side we have a convective entropy production. If we claim the Principle of Maximum Entropy Production, we should also claim the *Principle of Maximum Dissipation* (from the 4th Balance Theorem) as well as the *Principle of Maximum Heat Flux*. And that principle is how Malkus' and Veronis' result is sometimes known in literature. Unfortunately, their stability analysis was inadequate, which precluded them from describing an increased cell size with increased Rayleigh number. In his two papers (Malkus, 1954a,b) Malkus put forward the so called 'hypothesis of maximum heat transport', which is exactly equivalent to the principle of maximum entropy production.

An important result was obtained by Schlüter, Lortz, and Busse (1965). Using a stability analysis, they proved that steady state convection in an infinite layer must be in the form of two- dimensional rolls, which is confirmed by the vast majority of experiments. Moreover, they have found that two - dimensional rolls provide the maximal heat transport compared to any three- dimensional cells. Unfortunately, their analysis led them to the wrong conclusion regarding cell size uniqueness; they found that the cell size is not unique. Also, on their stability diagram they showed a stability area at values of the wave number larger than the critical wave number k_{cr}, which contradicts the experiments shown in Figure 1, where all curves are situated at k-values smaller than the critical wave number k_{cr}, that is at λ-values larger than the critical wavelength $\lambda_{cr} = 2$.

Another attempt to find criteria for the selection of k was made by Inoue and Ito (1984). They introduced and examined five criteriae for the selection of the wave number k:

1) Maximum mean Nusselt number;
2) Maximum velocity;
3) Maximum temperature amplitude;
4) Maximum entropy production;
5) Minimum generalized entropy production.

Because Inoue and Ito restricted their analysis to the trial functions from linear theory, they could not obtain theoretical results consistent with the experimental distribution $k(Ra)$. At the end of their work they wrote: "The analytical method of this paper does not produce an accurate estimation of the wave number."

6. **Approximate analytical solution for the W-model**

I will seek a solution of the model (4.1)-(4.8), satisfying the following conditions:

1) It has the same cell size behavior as the experimental curves $\lambda(Ra)$, $d\lambda/dRa > 0$;

2) It exactly satisfies the W-model (4.1)-(4.2), (4.4) globally, that is over a cell's volume, and it satisfies the boundary conditions (4.3), (4.5)-(4.8) locally;

3) It is unique.

89

Chapter IV. **Wave number selection**

Uniqueness is important because it is a characteristic of all experiments for Standard Convection Conditions. It has occasionally been maintained that non - unique convection motions have been observed; however, only in non - uniform fields of convection. Such non - unique motions are due to local non - uniformities of the temperature field, as they occur, e.g. at the rim of the fluid. Non - uniqueness of supercritical convection can be proven experimentally only by showing the existence of a steady uniform field of convection in the entire layer at a given supercritical Rayleigh number, and the existence of another steady uniform field of convection with a different cell-size in the same apparatus, the same fluid, at the same supercritical Rayleigh number. That has not been done so far.

Cell-size behavior λ(Ra) is even more important because it is the stumbling block between theories and experiment. While the latter is showing increased cell size at increased Ra, $d\lambda/dRa > 0$, the theories predict the opposite. This contradiction was noted in several publications, but its fundamental character for the development of convection theory was stressed in Koschmieder's book (1993). Comparison of the theory with experiments needs to be done on a similar basis. Experiments and theory must satisfy:

a) Steady state conditions;
b) Uniqueness of solution;
c) An infinite layer or a container with large Aspect Ratio;
d) Boundary conditions must be constant;
e) Fluid parameters (density, viscosity,...) must be constant.

If the listed conditions are satisfied, I will call the curves λ(Ra), or the inverse curves Ra($k = 2\pi/\lambda$) = Ra(k), obtained by experiment or by theory, ***Koschmieder's curves.*** The picture (Fig.4.1 above), presented in Koschmieder's book, (1993, page 92), is composed of several experimental curves $\lambda(Ra)$, all of which satisfy the mentioned conditions.

Let us build the approximate analytical solution for the case of solid boundaries and in parallel for the case of free boundaries.

I will add letters "S" and "F" to the corresponding equation numbers. To make it simple, yet retaining qualitative features of the exact solution, I use the following assumptions:

1) Approximation of the velocity V for solid boundaries:
$$V_z = (a/\pi)\cos(kx)\sin^2(\pi z), \quad V_x = -(a/k)\sin(kx)\sin(2\pi z), \qquad (6.1S)$$
and for free boundaries:
$$V_z = (a/\pi)\cos(kx)\sin(\pi z), \quad V_x = -(a/k)\sin(kx)\cos(\pi z), \qquad (6.1F)$$
The reason for the approximation (6.1S) is the experiments by Berge (1975), where he has shown that in moderately supercritical steady state convection "a single mode of constant wavelength is found for both V_x and V_z velocities".

The second reason for V_z in (6.1S) is suggested by linear theory $V_z = (a/\pi)\cos(kx)w(\pi z)$, where rather complicated exact $w(\pi z)$ is approximated by simple $\sin^2(\pi z)$, preserving the general shape, the symmetry with respect to $z = \frac{1}{2}$, and the boundary conditions. The free boundaries V_z in (6.1F) are taken exactly from the linear theory.

After V_z is defined, V_x formula is derived from the continuity equation $\mathrm{div}V \equiv \partial_x V_x + \partial_z V_z = 0$.

2) Approximation of the temperature T for solid boundaries:
$$T = b\cos(kx)\sin^2(\pi z) + c\sin(2\pi z), \qquad (6.2S)$$

and for free boundaries $T=b\cos(kx)\sin(\pi z)+c\sin(2\pi z)$. (6.2F)

Here again the first term in (6.2) is suggested by linear theory. As for (6.2S), $\sin^2(\pi z)$ has its derivative value (that is heat flux) equal to zero at the boundaries. The last term $c\sin(2\pi z)$ provides equal fluxes dT/dz at the boundaries that are non - zero, which is necessary for steady state convection. It is that last term in (6.2) along with maximum entropy production that makes all the difference from the linear theory. Also, in relation to his mentioned experiment, Berge (1975) writes: "one can see a sinusoidal dependence of the temperature perturbation".

Thus, reasons 1) and 2) were made for choosing the approximations (6.1) - (6.2), where the constant amplitudes a, b, c are unknown up to now. To find three amplitudes I need three equations. One of them is provided by Chandrasekhar's Balance Theorem, from which we have $a=b\mathrm{Ra}f(k)$, where $f(k)$ will be specified below. The second equation can be obtained by multiplying the heat equation (1.2) by $\cos(kx)\sin^2(\pi z)$ (in free boundaries by $\cos(kx)\sin(\pi z)$) and integrating over the cell volume

$$\Omega = \{-\pi/k \le x \le \pi/k, \quad 0 \le z \le 1\}.$$

The third equation can be obtained by multiplying the heat equation (1.2) by $\sin(2\pi z)$ and integrating over the cell volume (Galerkin method).

The second equation is especially interesting, because it provides the Neutral Curve without applying a stability analysis. Indeed, the second equation for solid and free boundaries is

$$9k^6 + 36k^4\pi^2 + 80k^2\pi^4 + 64\pi^6 - 9k^2\text{Ra} - 12ck^2\pi\text{PrRa} = 0. \qquad (6.3\text{S})$$

$$k^6 + 3k^4\pi^2 + 3k^2\pi^4 + \pi^6 - k^2\text{Ra} - ck^2\pi\text{PrRa} = 0. \qquad (6.3\text{F})$$

As all amplitudes must vanish, along the Neutral Curve, setting $c = 0$ in (6.3), one can get the equation of the Neutral Curve:

$$\text{Ra}_0(k) = (3k^2 + 4\pi^2)(3k^4 + 8k^2\pi^2 + 16\pi^4)/(9k^2). \qquad (6.4\text{S})$$

$$\text{Ra}_0(k) = (k^2 + \pi^2)^3/k^2. \qquad (6.4\text{F})$$

A simple, but not very precise Neutral Curve for solid boundaries is shown in Figure 6.1 along with the exact curve of Pellew and Southwell (1940). Another way to find the Neutral Curve is through the use of the equation

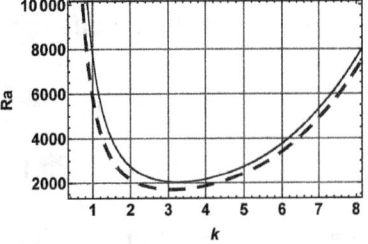

$$W(k, \text{Ra}) = \int (\nabla T)^2 d\Omega = 0, \qquad \textbf{(6.5)}$$

Fig.6.1. Neutral curves: dashed -exact, and solid-approximate. Solid boundaries.

The use of (6.5) should precede finding the amplitudes a, b, c. Solving the three equations, one is the Chandrasekhar's Balance and two are from the Galerkin method, with three unknown amplitudes a, b, c one obtains results (6S for solids, 6F for free, 6 for both boundaries):

$$a = (\text{Ra} - \text{Ra}_0(k))^{1/2}F(k)/\text{Pr}, \quad b = (\text{Ra} - \text{Ra}_0(k))^{1/2}G(k)/(k\text{RaPr}),$$
$$c = (\text{Ra} - \text{Ra}_0(k))H(k)/(\text{RaPr}), \qquad (6.6)$$

$$f(k) = (3k^4 + 8k^2\pi^2 + 16\pi^4)^{1/2}, \qquad F(k) = 3\sqrt{2}/f(k),$$
$$G(k) = f(k)\sqrt{2}, \qquad H(k) = 3/(4\pi), \qquad (6.6\text{S})$$

$$F(k) = 2\sqrt{2}/(k^2 + \pi^2), \quad G(k) = 2\sqrt{2}(k^2 + \pi^2), \quad H(k) = 1/\pi, \quad (6.6F)$$

To illustrate how entropy production determines the approximate Neutral Curve, I insert our temperature (6.2) into (6.5), and after integration we get:

$$W = 3b^2k\pi/8 + \pi^3(b^2 + 8c^2)/(2k). \tag{6.7S}$$

$$W = (8c^2\pi^3 + \pi(b^2 + \pi^2))/(2k). \tag{6.7F}$$

Substituting b and c from (6.6) into (6.7) gives:

$$W(k,\mathrm{Ra}) = \pi((3k^2 + 4\pi^2)(3k^4 + 8k^2\pi^2 + 16\pi^4) - 9k^2\mathrm{Ra})/(4k^2\mathrm{Pr}^2\mathrm{Ra}^2), \tag{6.8S}$$

$$W(k,\mathrm{Ra}) = (-4\pi(k^2 + \pi^2)^3 + 4\pi k^2\mathrm{Ra})/(k^3\mathrm{Pr}^2\mathrm{Ra}). \tag{6.8F}$$

$W = 0$ gives the Neutral Curves (6.4S), (6.4F). We can rewrite the entropy production W, using the Neutral Curve $\mathrm{Ra}_0(k)$:

$$W = 9\pi(\mathrm{Ra} - \mathrm{Ra}_0(k))/(4k\mathrm{RaPr}^2). \tag{6.9S}$$

$$W = 4\pi(\mathrm{Ra} - \mathrm{Ra}_0(k))/(k\mathrm{RaPr}^2). \tag{6.9F}$$

The entropy production $W(k,\mathrm{Ra})$ is presented by (6.9) in general form. W increases when k decreases, so the maximum of W is shifted to smaller k when Ra increases.

And this means that the cell size $\lambda = 2\pi/k$ is growing along with Ra.

The surface $W=W(k,\mathrm{Ra})$ in Figure 6.2 is entropy production.

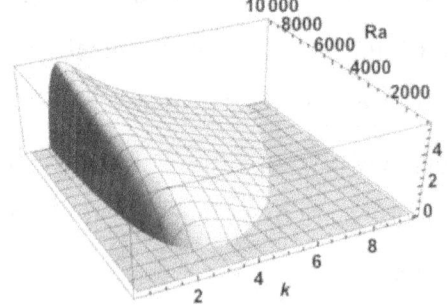

The curve $\mathrm{Ra}_{max}(k)$ in (7.2) is a projection of the summit

Fig.6.2. The surface of entropy production $W=W(k,\mathrm{Ra})$. Pr=0.7(Air), Solid boundaries.

of the mountain range of $W(k,\mathrm{Ra})$ onto the plane (k,Ra).

7. Cell size and maximum entropy production

First, I will consider the maximum entropy production as a function $\max W(k,\text{Ra}) = \text{Ra}_{max}(k)$, and then I will convert it into $\lambda(\text{Ra})$.

All I have to do is to solve the equation

$$\partial W(k,\text{Ra})/\partial k = 0 . \tag{7.1}$$

From (6.9), (7.1) I obtain the same equation for solid and free boundaries, but with different Neutral Curves $\text{Ra}_0(k)$:

$$\text{Ra}_{max}(k) = \text{Ra}_0(k) - k(\partial \text{Ra}_0/\partial k) . \tag{7.2}$$

$$\text{Ra}_{max}(k) = 80\pi^4/9 - 4\pi^2 k^2 - 3k^4 + 64\pi^6/(3k^2) , \tag{7.3S}$$

$$\text{Ra}_{max}(k) = - 3(k^2 - \pi^2)(k^2 + \pi^2)^2/k^2 , \tag{7.3F}$$

From (7.3) it is easy to see that

$$\partial \text{Ra}_{max}(k)/\partial k < 0. \tag{7.4}$$

$\text{Ra}_{max}(k)$ increases when k decreases (or $\lambda = 2\pi/k$ increases) from its critical value k_{cr}. $\text{Ra} = \text{Ra}_{max}(k)$ is shown in Fig.7.1 along with the Neutral Curve $\text{Ra}_0(k)$.

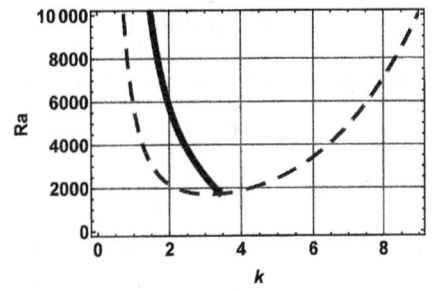

Fig.7.1. The thick line $\text{Ra}_{max}(k)$ shows the cell size behavior. The dashed curve is the Neutral Curve.   Solid boundaries.

To make Ra_{max} more comparable with the experimental curve, such as for air in Fig.4.1, one may recalculate the graphs from Fig.7.1 to a different scaling, that is to $\lambda(\text{Ra})$ as in Fig.7.2.

Both graphs have the beginning of the curve a little shifted from the actual critical point (k_{cr}, Ra$_{cr}$) due to the approximate solution (6.1), (6.2). Those shifts will disappear when we do all the calculations for the case of free boundaries.

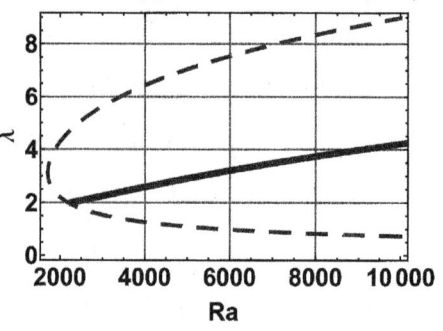

Figure 7.2 shows that λ(Ra) has reasonable values compared to the experimental curves in Fig.1.1.

Fig.7.2. λ_{max}(Ra) increases within the Neutral curve. Air.

Repeating the procedure outlined for the solid boundaries, one can get the surface (6.8F) for free boundaries, from which the classical Neutral Curve is obtained at $W = 0$:

$$Ra_0(k) = (k^2 + \pi^2)^3/k^2 . \qquad (7.5)$$

The entropy Production surface for the W-model with free boundaries and Pr=0.7 is shown in Fig.7.3. It is similar to the case with solid boundaries (Fig.6.2).

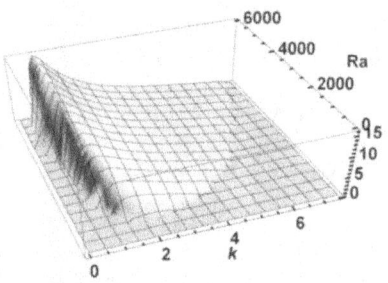

Fig.7.3. Free boundaries. Entropy production surface W. Steady state.

These graphs confirm the experimental observation:

mental observation:

1) The cell size increases with Ra monotonically;
2) The cell size depends uniquely on Ra.

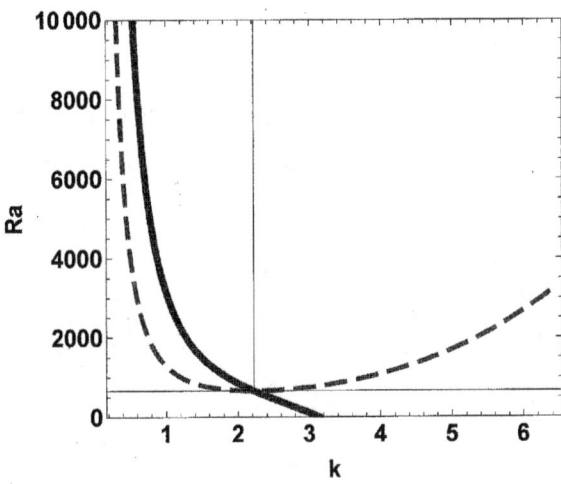

Fig.7.4. Free boundaries. The thick line Ra$_{max}$(k) shows the cell-size behavior. The dashed curve is the Neutral Curve. Vertical and horizontal lines intersect at the critical (k_{cr}, Ra$_{cr}$).

8. Landau's Law

In this section I will present Landau's Square Root Law and its extension derived from the approximate solution of the W-model for solid and free boundaries.

The *Landau's Square Root Law* states that above the critical regime up to a certain point, amplitude *a* of the fluid velocity increases as a square root of the difference between the current flow and its critical value, which Landau considers constant:

$$a \sim (Ra - Ra_{cr})^{1/2}. \tag{8.1}$$

In the W-model, the value of the amplitude is not a constant, but a function of Ra$_0$(k), which is the Neutral Curve. Only the minimum of that function is constant, Ra$_{cr}$ = Ra$_0$(k_{cr}), corresponding to the critical value. To show Landau's Law and its extension, I rewrite the solution of the W-model (6.6):

$$a =(Ra - Ra_0(k))^{1/2}F(k)/Pr, \quad b =(Ra - Ra_0(k))^{1/2}G(k)/(kRaPr),$$
$$c=(Ra - Ra_0(k))H(k)/(RaPr), \qquad (8.2)$$

Landau derived his Square Root Law for general hydrodynamic flow. He did not consider convective flow. In (8.2) only the velocity amplitude a follows the Square Root Law. The flux amplitudes ab and c are linear functions of Ra - Ra_0 divided by Ra.

Although the W-model is made for the moderately supercritical regime, nevertheless it is interesting to see the behavior of the amplitudes for the very large Ra $->\infty$:

$$a \rightarrow \infty, \ ab \rightarrow constant, \ c \rightarrow constant, \ W \rightarrow constant.$$

This behavior is the same for solid and free boundaries.

9. Concluding remarks

A simple W-model of steady state convection was presented in this chapter. The model was based on the Maximum Principle that can be stated in the following equivalent forms:

1) The Principle of Maximum Entropy Production,
2) The Principle of Maximum Heat Flux (Second Balance Theorem),
3) The Principle of Maximum Dissipation (Fourth Balance Theorem).

The model confirms the main features of many independent experiments about cell-size behavior in steady state convection at Standard Convection Conditions:

1) The cell size monotonically increases with increased Ra;

2) The cell size is a unique function of Ra.

REFERENCES

Bénard, H. 1900. Les tourbillions cellulaires dans une nappe liquid. *Rev. Gen. Sciences Pure Appl.* **11,** 1261-1271, 1309-1328.

Behringer, R.P. 1985, Rayleigh-Benard convection and turbulence in liquid helium. *Reviews of Modern Physics*, Vol.57, No.3, pp.657-687.

Bergé, P. 1975. Rayleigh - Benard instability: Experimental findings obtained by light scattering and other optical methods.
In *Fluctuations, Instabilities and Phase Transitions*, pp.323 - 352, ed. T.Riste. Plenum Press, New York.

Bodenschatz E., Pesch W., Ahlers G., Recent development in Rayleigh - Benard convection, *Annu.Rev.Fluid Mech.* 2000, 32:709 - 777.

Buell, Jeffrey C. and Catton, Ivan 1986, Wavenumber selection in large-amplitude axisymmetric convection, *Phys. Liquids* **29** (1), 23-30.

Busse, F.H., 1967. On the stability of two - dimensional convection in a layer heated from below. *J. Math. Phys.,* **46,** №2,140 - 150.

Busse, F.H., & Whitehead J.A., 1971. Instabilities of convection rolls in a high Prandtl number fluid. *J. Fluid Mech.,* **47,** 305 - 420.

Catton, Ivan 1988, Wavenumber selection in Benard convection, *Journal of Heat Transfer,* **110**, 1154-1162.

Chandrasekhar, S. 1981 Hydrodynamic and Hydromagnetic Stability, Dover Publications.

Chen, M.M., & Whitehead J.A., 1968, Evolution of two-dimensional periodic Rayleigh convection cells of arbitrary wave-number. *J. Fluid Mech.* **31**, 1 - 15.

Davis, S.H. and Segel. L.A., 1968, Effect of curvature and property variation on cellular convection. *Phys. Fluids,* **11**, 470-6. [p.41]

Drazin, P.G. and Reid, W.H. 1984, Hydrodynamic Stability, Cambridge University Press.

Farhadieh, R., & Tankin, R.S., 1973. Interferometric study of two - dimensional Benard convection cells. *J. Fluid Mech.* **66,** 739 - 753.

Foster, T.D., 1969, The effect of initial conditions and lateral boundary conditions on convection, *J. Fluid Mech.* **37**(1)**,** 81 - 94.

REFERENCES

Inoue, Y., & Ito, R. 1983. Analysis of Bénard convection by the energy - integral method. *Int. Chem. Engr.* **24,** 311 - 320.

Гершуни Г.З. и Жуховицкий Е.М. Конвективная устойчивость несжимаемой жидкости, Изд.Наука, Москва,1972г.

Gershuni,G.Z. and Zhukhovitskii, E.M., 1976 Convective Stability of Incompressible Fluids, (Translated from Joseph, Daniel D., 1976 *Stability of Fluid Motion.* Springer-Verlag. Russian edition of 1972), Israel Program for Scientific Translations, Jerusalem.

Koschmieder E.L., 1993. Benard Cells and Taylor Vortices, Cambridge University Press, p.75.

Koschmieder, E.L., & Pallas, S.G. 1973. Heat transfer through a shallow, horizontal convecting fluid layer. *Int. J. Heat Mass Transfer* **17,** 991 – 1002.

Krishnamurti, R. I970. On the transition to turbulent convection. Part 1.The transition from two- to three- dimensional flow. *J. Fluid Mech.* **42,** 295 – 306.

Landau, L.D., & Lifshitz, E.M. 1986. Stability of steady flow. In *Fluid Mechanics*, 2nd edition, pp.95 – 98, Pergamon Press, New York, U.S.A.

Malkus, W.V.R. 1954a. Discrete transitions in turbulent convection. *Proc. Roy. Soc.* **A, 225,** 185 – 195;

Malkus, W.V.R. 1954b. The heat transport and spectrum of thermal turbulence. *Proc. Roy. Soc.* **A, 225,** 196 – 212;

Malkus, W.V.R. and Veronis, G. 1957. Finite amplitude cellular convection. *J. Fluid Mech.* **4**, 225 – 260.

Martinet, B., Haldenwang, P., Labrosse,G., Payan, J. – C.,& Payan,R. 1983. Selection des structures dans l'stabilite de Rayleigh – Benard. *Compt. Rend. Ser.2* **299,** 755 – 757.

Pellew, A. & Southwell, R.V., 1940. On maintained convective motions in a fluid heated from below. *Proc. Roy. Soc.* London **A, 176,** 312 – 343.

Rayleigh, Lord. (1916). On convective currents in a horizontal layer of fluid, when the higher temperature is on the under side. *Phil. Mag.***32**, 529. Collected papers, **6**, 432.

Russel, "On obtaining solutions to the Navier-Stokes equations with automatic digital computer", Oxford, 1962

(Sorokin) Сорокин В.С. *Вариационный метод в теории конвекции,* Прикладная Механика и Математика, 1953, Том 17, №1, 39.

REFERENCES

Schlüter, A., Lortz, D., and Busse, F. 1965. On the stability of steady finite amplitude convection. *J. Fluid Mech.* **23,** 129 – 143.

Silveston,P.L. 1957. Wärmedurchgang in waagerehten Flüssigkeitsschichten. *Forsch. Ing. Wes.* **24,** 29 – 32, 59 – 69.

Swenson, R. 1987. Emergence and the principle of maximum entropy production: Multi – level system Meeting of the International Society for General Systems Research, theory, evolution, and non – equilibrium thermodynamics. Proceedings of the 32nd Annual Meeting of the International Society for General Systems Research, 32: http://www.lawofmaximumentropyproduction.com/

Ukhovskii M.P., Yudovich V.I. 1963, On the equations of steady state convection., *Journal of Applied Mathematics and Mechanics,* **27,** №2, 1963, 295-... .

Willis, G.E., Deardorff, J.W., & Somerville, R.C.J. 1972. Roll – diameter dependence in Rayleigh convection and its effect upon the heat flux. *J. Fluid Mech.* **54,** 351 – 366.

Yudovich V.I. 1966, On the onset of convection., *Journal of Applied Mathematics and Mechanics,* **30,** 1193.

Yudovich V.I. 1967, Free convection and bifurkation, *Prikl. Mat Mech,* **31**(1), 101-111.

ADDENDUM

In this addendum I continue to use steady state and 2D-geometry with horizontal axis x and with vertical axis z.

1. Equations and boundary conditions

With Ra, Pr, ρ_0, **g**, γ, α, u as 7 constant parameters, the main equations and boundary conditions are (steady state: $\partial V/\partial t = 0$, $\partial T/\partial t = 0$):

$$(V\nabla)V = -\nabla P + \Delta V + \text{Ra}T\gamma, \tag{a.1}$$

$$(V\nabla)V = -\nabla P/\rho_0 + v\Delta V + g(-\alpha T + V^2/(2u^2)), \tag{b.1}$$

$$\text{Pr}V\nabla T = \Delta T + V_z, \tag{a.2}$$

$$\text{div}V = 0, \tag{a.3}$$

$$T = P = 0, \qquad\qquad z = 0 \text{ or } 1, \tag{a.4}$$

$$V_z = \partial V_z/\partial z = 0, \qquad\qquad z = 0 \text{ or } 1. \tag{a.5}$$

As a momentum equation I used (a.1) in Chapters I and IV and I used (b.1) in Chapters II and III.

In this Addendum all integrals will be evaluated in a similar manner. The majority of integrals will be reduced to zero by zero conditions at the surface, (a.4) –(a.5). The main instrument will be the Divergence theorem[1] on the reduction of the volume integral to the surface integral.

---

[1]Three men proved independently this theorem: Lagrange in 1762, Gauss in 1813, Ostrogradsky in 1826. In literature the Theorem has three names: Gauss theorem, Gauss-Ostrogradsky theorem, and Divergence theorem.
In my opinion the proper name for the theorem should be either first person (Lagrange), or all three of them or Divergence theorem.

I use the surface Σ containing the volume Ω; $n=\{n_x, n_z\}$ is a unit vector along the outward normal to Σ.

On the upper and lower boundaries $n = \{0, \pm1\}$. On the vertical boundaries there is a similar situation. In the Introduction I have defined the volume $\Omega = \{ -\pi/k \leq x \leq \pi/k, \ 0 \leq z \leq 1\}$, surface $\Sigma(z=0, \ z=1\text{or } d)$, cells and their diameters $\lambda = 2\pi/k$.

I also use equation (a.3) extensively, which is a consequence of incompressibility. The integration by parts will be the 3rd element in our arsenal. Along with vectors I use tensor representation

$$\partial_I = \partial x_i \, , \ \partial_i V_k = \partial V_k/\partial x_i$$

and summation assumed over repeating indices.

2. Integrals

(Please note that there is a difference between equations numbered with A... and those numbered with a...)

We begin with $\qquad \int V d\Omega = 0.$ $\hfill$ (A.1)

The reason for this is the steady state, at which the volume Ω (be it infinite layer or closed container) is completely filled with straight convective cells. Then there are two cases.

The case of an infinite layer.

A cell consists of two rolls. If a roll is divided into two halves by a vertical plane going through the roll's axis, then the velocity V has an opposite sign in those two halves. This means the flux of the liquid on the left side is equal to the flux on the right side with an opposite sign. Thus, the total flux of each roll is zero as in (A.1)

The case of a closed container.

In the case of the closed container, the argument for (A.1) is even simpler. Indeed, if the integral in (A.1) isn't zero, it would mean that total flux is leaving (or coming in) the container, which is impossible. Thus, (A1) is valid and

$$\int V_z d\Omega = 0. \tag{A.1z}$$

Integrals for the heat equation

$$\mathrm{Pr} V \nabla T = \Delta T + V_z \tag{A.2}$$

Integrating it term by term I use $V \nabla T = div(VT) - T div V$.

Here the second term is zero because of (a.3). The integral of the first term reduces to the surface integral $\int div(VT) d\Omega = \int VT d\Sigma = 0$ due to (a.4). Here Σ is a surface containing the volume Ω. Now we are left with

$$0 = \int \Delta T d\Omega . \tag{A.3}$$

Let's simplify it. $\int \Delta T \, d\Omega = \int div(\nabla T) d\Omega = \int \nabla T d\Sigma =$

$$= \int [(\partial T/\partial z)_{z=1}] dx - \int [(\partial T/\partial z)_{z=0}] dx = Q_c - Q_h = \Delta Q = 0, \tag{A.4}$$

where ΔQ is the difference between heat fluxes exiting, Q_c, and entering, Q_h, the fluid's layer. $\quad 0 = \partial_t \int T d\Omega = \Delta Q.$

The important physical consequence is that ***during the steady state, heat fluxes entering and exiting are equal.***

Integrals for the momentum equation

$$(V \nabla) V = -\nabla P / \rho_0 + v \Delta V + g(-\alpha T + V^2/(2u^2))$$

Now I integrate the momentum equation (b.1).

Its left side can be treated by the use of tensors:

$$(V \nabla) V = V_i \partial_i V_k = \partial_i (V_i V_k) - V_k \partial_i V_i = \partial_i (V_i V_k) - V_k \, div V = \partial_i (V_i V_k).$$

$$div V = 0. \qquad (V \nabla) V = \partial_i (V_i V_k).$$

Then reducing the volume integral to the surface integral according to the Divergence theorem , we obtain:

$\int (V\boldsymbol{\nabla}) V d\Omega = \int \partial_i(V_iV_k) d\Omega = \int n_i(V_iV_k) d\Sigma$, where Σ is a surface containing the volume Ω, and $\boldsymbol{n}=\{n_x, n_z\}$ is a unit vector along the outward normal to Σ.

On the upper and lower boundary $\boldsymbol{n} = \{0, \pm1\}$ and $V_z = 0$.

On the vertical boundary there is a similar situation, therefore

$$\int (V\boldsymbol{\nabla}) V d\Omega = \int n_i(V_iV_k) d\Sigma = 0.$$

Thus, $\qquad \int (V\boldsymbol{\nabla}) V d\Omega = 0, \qquad\qquad\qquad\qquad (A.5)$

$$0 = -\boldsymbol{\nabla}P/\rho_0 + v\Delta V + g(-\alpha T + V^2/(2u^2)).$$

Let's treat the other terms. Note that (b.1) is a system of two equations: one for V_x, one for V_z. This doesn't result in anything interesting for the first equation with V_x, but it is the second equation we are interested:

$$-\int \boldsymbol{\nabla}P d\Omega + \eta\int \Delta V_z d\Omega + g\rho_0\int (\alpha T - V^2/u^2) d\Omega = 0.$$

The integral of $\boldsymbol{\nabla}P$ can be reduced due to (a.4) to

$$\int \boldsymbol{\nabla}P d\Omega = \int Pn_z d\Sigma = 0 . \qquad\qquad\qquad\qquad (A.6)$$

For integral of ΔV_z one can get the result below due to (a.5):

$$\int \Delta V_z d\Omega = \int [(\partial V_z/\partial z)_{z=1}] dx - \int [(\partial V_z/\partial z)_{z=0}] dx = 0. \qquad (A.7)$$

In the Momentum equation (b.1) all but the last terms are zero, so

$$\int (\alpha T - V^2/(2u^2)) d\Omega = 0. \qquad\qquad\qquad\qquad (A.8)$$

The kinetic energy equation

Now we'll integrate kinetic energy equation (I.5.1) term by term. For the steady state it is

$$-\rho_0 V(V\boldsymbol{\nabla})V - V\boldsymbol{\nabla}P + \eta V\Delta V - g\rho_0(\alpha VT - VV^2/(2u^2)) = 0 . \qquad (A.9)$$

First, I prove $V(V\boldsymbol{\nabla})V = div(VV^2/2). \qquad\qquad\qquad\qquad (A.10)$

$$V(V\nabla)V = V_i V_k \partial_k V_i = V_k (\partial_k V^2_i/2) = \partial_k(V_k V^2_i/2) - (V^2_i/2)\partial_k V_k =$$
$$\text{div}(VV^2/2) - (V^2_i/2)\,\text{div}V = \text{div}(VV^2/2) - 0.$$

Integrating (A.9) and applying the Divergent theorem, we get

$$\int V(V\nabla)V d\Omega = \int \text{div}(VV^2/2)\,d\Omega = 0 . \qquad (A.11)$$

The next term is $V\nabla P = \text{div}(VP) - P\text{div}V,$ $\text{div}V = 0,$

$$\int (V\nabla P)d\Omega = \int \text{div}(VP)d\Omega = \int(VP)d\Sigma = 0 .$$

$$\int (V\nabla P)d\Omega = 0. \qquad (A.12)$$

The next term is $V\Delta V.$

According to Landau and Lifshitz (1986, §15), momentum flux is

$$\sigma_{ik} = \eta(\partial_k V_i + \partial_i V_k) \qquad (A.13)$$

and the kinetic energy equation has a term $\sigma_{ik}\partial_k V_i$, which is in fact $(-V\Delta V),$ and the dissipative function defined as $D = \sigma_{ik}\partial_k V_i$.

Thus $\eta V\Delta V = -\sigma_{ik}\partial_k V_i = -\eta(\partial_k V_i + \partial_i V_k)\,\partial_k V_i$. Due to the symmetry, the second factor, $\partial_k V_i$, is replaced by $(\partial_k V_i + \partial_i V_k)/2$, therefore

$$D = \eta(\partial_k V_i + \partial_i V_k)^2/2 . \qquad (A.14)$$

The last term in (A.9) is $g\rho_0(\alpha V_z T - V_z V^2/u^2).$

$$\int V_z T d\Omega = B - \text{vertical heat flux.} \qquad (A.15)$$

$\int \rho_0 V_z V^2 d\Omega = 0$ – vertical weighted mass flux. If it isn't zero, fluid would go outside of volume Ω.

Taking non-zero terms from kinetic equation (A.9) after integration we have a Chandrasekhar's Balance Theorem

$$g\rho_0\alpha\int V_z T d\Omega - \int D d\Omega = 0. \qquad (A.16)$$